FROM THE GROUND UP

Lewis Mumford was born in Flushing, Long Island, in 1895, and was educated at the College of the City of New York, Columbia University, and the New School. Patrick Geddes's work on city development inspired him to become an active student of cities; he first surveyed in detail the entire metropolitan area of New York, and then worked and studied in other cities, including Pittsburgh, Boston, and London. His first book, *The Story of Utopias,* was published in 1922. This was followed in 1924 by *Sticks and Stones.* Since that time he has published many books on buildings and cities, including *The Brown Decades* (1931), *The South in Architecture* (1941), *City Development* (1945), *From the Ground Up* (1956), *The City in History* (1961), and *The Highway and the City* (1963). The four volumes of his monumental series *The Renewal of Life* are: *Technics and Civilization* (1934), *The Culture of Cities* (1938), *The Condition of Man* (1944), and *The Conduct of Life* (1951). From 1951 to 1956 Mr. Mumford was Visiting Professor at the School of Fine Arts, University of Pennsylvania, and from 1957 to 1960 he was visiting Bemis Professor at the Massachusetts Institute of Technology. In 1957 he received the gold medal of the Town Planning Institute, and in 1961 he was awarded the Royal Gold Medal for Architecture by the Royal Institute of British Architects. He became an honorary associate of the R.I.B.A. in 1942: he is likewise an honorary member of the Town Planning Institute (London), the American Institute of Planners, the Town Planning Institute of Canada, and the American Institute of Architects.

From the Ground Up

OBSERVATIONS ON
CONTEMPORARY ARCHITECTURE,
HOUSING, HIGHWAY BUILDING,
AND CIVIC DESIGN

By Lewis Mumford

A Harvest Book
HARCOURT BRACE JOVANOVICH, INC.
NEW YORK

NA
2560
M8

Preface

In The Sky Line department of *The New Yorker,* a little
while ago, I published a series of articles on contemporary building, highway planning, and civic design, called
"The Roaring Traffic's Boom." The countrywide demands for reprints of that series made me at last take
seriously a project I had long kept in abeyance: the
putting together of my Sky Line criticisms in a book.
Though these reviews are confined to New York, the
issues they raise are universal ones; and on the understanding of these issues by the ordinary citizen, as well
as by the architect, the builder, the municipal administrator, and the financier the health of our whole civilization depends. Shall we produce order or chaos? spa-
ciousness or congestion? aesthetic delight or depression?
townscapes and landscapes designed for living or cells
and prison blocks for automatons? these are some of
the questions this book raises and tries, in some degree,
to answer.

In finally consenting to bring these pieces together
I have made a rigorous selection, passing over criticisms
of purely local interest, and cheerfully consigning to
oblivion all my contributions between 1932 and 1947.
The Sky Line department itself was one of the many
brilliant contributions that the late H. W. Ross, the first
editor of *The New Yorker,* made to contemporary jour-
nalism. My predecessor, "T-Square," happily set a

level for forthright criticism that Mr. Ross, who regarded this architectural department as a public service, consistently upheld, even though it cost the magazine, before I joined it, a lawsuit. In publishing these pieces I wish to record affectionately my personal debt to Mr. Ross: his persistent interest, his passionate grasp of detail, and his unwavering loyalty in the face of attack were no small contributions to the department. My gratitude to other members of *The New Yorker* staff extends clear through to the checkers, those anonymous marvels of patience, vigilance, and exact scholarship, who rectify errors of fact before they can appear painfully in print, and who supplement my occasionally sketchy descriptions with all the gruesome details. But most especially my gratitude goes to two editors, Mr. Rogers Whitaker, whose cooperation is often close to collaboration, and Mr. William Shawn, who presides over and maintains *The New Yorker* traditions that Ross established. Perhaps, finally, I should give a friendly nod of thanks to the architects whose work I have reviewed: sometimes they have rewarded me in words, but sometimes by their silence.

                                                           L. M.

Contents

THE ROARING TRAFFIC'S BOOM

"MAKE NO LITTLE PLANS"

ONE

From Utopia Parkway Turn East

At Fresh Meadows, in Queens, the New York Life Insurance Company has built a great housing project. When the news gets around, people will be making Sunday excursions there to see for themselves, enviously, how pleasant the rest of New York might be if our public officials and our investors ever acquired the wisdom that has guided the New York Life Insurance Company in this venture.

"Great" is a word I use sparingly, especially about housing projects, but when I first saw the plans for Fresh Meadows, I had a hunch that they would be well above the average in many ways. I did have misgivings about the advisability of the two huge apartment houses, laid out in the form of crosses with double bars, that dominate the otherwise modest sky line of the project; and now that the buildings are up, I feel that my misgivings about them were justified. But in almost every other respect, the owners and the architects have exceeded my most sanguine expectations.

This is the only large-scale project I know of that rivals Baldwin Hills Village, in Los Angeles, and it is the only one in this part of the country, except Greenbelt, in Maryland, that presents a detailed view of what the residential neighborhoods of our cities would be like if they were planned not merely with a view to creating a safe, long-term investment but also to promote the comfort, the joy, and the equability of their inhabitants. Both the design and the execution of this development deserve diligent scrutiny. The Metropolitan Life Insurance Company's Stuyvesant Town, in lower Manhattan, and the New York City Housing Authority's projects are painful lessons in how not to rebuild New York. Fresh Meadows is a fine antidote. And it is a far too concrete and practical demonstration for the Authority's Mr. Robert Moses to dismiss it as the idle dream of long-haired theorists.

Fresh Meadows, like Stuyvesant Town, is a private real-estate development, but, unlike Stuyvesant Town, it was built without public aid of any sort, as a sound and profitable investment for the New York Life's capital. The life-insurance companies of America have billions of dollars to invest; they invested over eleven billions last year, and the amount available should increase steadily. It is obviously important to all of us that some of this money should be invested not in the slums of midtown Park Avenue and other blighted districts, which remain profitable only as long as they remain congested, but in buildings, in both old and new urban areas, that will not turn sour so quickly. And for the insurance companies it is a matter of self-interest to improve our living quarters, since adding to the length of life of their policy-holders increases their profits. Quarters that tempt people to early marriage help, too, since married men have a longer life expectancy.

Because Fresh Meadows has been built by a big corporation, it has had the advantage of large-scale organization all the way, from the acquiring of the large tract on which it stands to the building of a central heating plant; from using a panel of life-insurance officials to select the eighteen physicians and six dentists who serve the community to employing the architec-

tural firm of Voorhees, Walker, Foley & Smith to
work with the New York Life's own architect, G. Har-
mon Gurney, in planning this complex neighborhood project literally from the ground up. Piecemeal
building by small investors simply cannot achieve the
economies or create the collective order and beauty
that a big operation can. The great early successes in
town planning and housing, such as the magnificent
eighteenth-century London squares of Bloomsbury
and Mayfair, were the work of the Duke of Bedford
and other enlightened landed proprietors. The New
York Life Insurance Company has given its architects
a chance to show how humane and attractive a modern community can be if the designer's imagination can
be applied not to isolated buildings but to the inter-
relationship of people, trees, greens, parks, streets, and
buildings, so that they become an organic unity.
Though only a few structures in Fresh Meadows,
among them the Nursery Center and Bloomingdale's
branch department store, are conspicuously handsome
in themselves, the community as a whole is probably
the best-looking piece of architecture in the metropolitan area, for it presents a series of architectural com-
positions, executed with a varied play of light and
shade, of masses and volumes, and of color, that have
all but disappeared from the urban architect's reper-
toire. Apart from those two dominating thirteen-story
apartment houses, the human scale (one- to three-
story buildings) has everywhere been maintained, and
the aesthetic qualities are balanced by human qualities;
in a community carpeted from end to end with lawns,
I could not find, except on newly seeded patches
around the skyscrapers, a single keep-off-the-grass
sign.

Fresh Meadows lies in the heart of Queens, to Manhattanites an unholy distance (fifty or sixty minutes by

subway and bus) from the center of things, and the approach is not wholly lovely. On the last lap of the outward journey by bus to the main entrance, at 188th Street and Horace Harding Boulevard, one passes through suburban streets rather like the Flatbush of forty years ago, filled with houses whose awkward efforts at domesticity have now been gracefully buried in foliage. This nostalgic bit of suburbia thins out into areas blighted by the rawest kind of post-Second-World-War private-housing enterprise, and these in turn are succeeded by open spaces rank with weeds. But before one reaches Fresh Meadows, one comes upon omens that portend something better. One travels for quite a distance along Utopia Parkway; then, turning east, one passes a Utopia Cleaners and, a little later, encounters a Utopia Auto Laundry. Thus one is prepared for splendor and glory. The first glimpse of the new community, if one gets off the bus at 186th Street and Horace Harding Boulevard, is a shock; the two big apartment houses, looking as vast as those in Stuyvesant Town, crown the middle distance.

But as one nears the main entrance, one finds that order and comeliness and charm pervade the design. One passes a series of commercial buildings of various heights and widths, which begin in a modest way to give the true architectural measure of the scene: first a one-story food market with wide windows; then a spacious brick bank, taller than the market, with a huge recessed panel of oblong glass panes across the whole front; then a lower drugstore. At the corner, there is a sudden change in scale and amenity: what in most communities would be an expensive commercial "busy corner" has no buildings at all. Instead, there is a small plaza, with trees and benches on the right, and a generous park, with more trees and more benches, in front of the big Bloomingdale store on the

left. This fine generosity with open spaces character-
izes the whole plan.

The deeper one penetrates into Fresh Meadows, the
more favorable the impression, for the architects present one with a series of urban vistas rare in a modern
American community—of short, curving streets, of
long, open greens and buildings beyond, of plentiful
verdure against a restful background of brick walls, of
wide windows and great pools of domestic quiet behind the long, irregular, widely spaced rows of three-
story apartment houses and two-story dwellings.

The Fresh Meadows housing is grouped around a
central open area, almost twenty acres, of rolling terrain. Most of the site was once a golf course, and this
central area still contains the clubhouse, which served
as headquarters during construction and will eventually vanish. The two tall apartment houses, a garage,
and the nursery school are the only permanent encroachments on this space; no disposition has yet
been made of the rest of it, including the land that will
be available when the clubhouse is demolished. To
allow this extra margin of space was prudent foresight;
planning must never be so tight that it leaves no room
to embrace the unexpected. I hasten to suggest, for
one thing, a small, well-equipped nursing home and
maternity center, so that hospitalization for all but
major physical distress can be on the intimate, humane
scale of the everyday life of the community. Large-
scale "warehousing of disease," to use the phrase of a
famous Victorian doctor, is one of the most barbarous
characteristics of our civilization, and the New York
Life is now in an admirable position to provide its tenants with something far better in a field of endeavor
so close to the corporation's vital interest.

The project covers a squarish tract of a hundred
and seventy acres, almost three times as large as Stuy-

vesant Town and one-fifth the area of Central Park. In
developing this block of land, the owners have gone
back to a reasonable urban density; despite the mass-
ing of humanity in the two big apartment units, they
have held down the population to seventeen families an
acre. Eleven thousand people are housed there, but
because of the high birth rate in the community, which
is occupied mainly by the families of young veterans,
it should presently be bursting at the seams, and the
management has already discovered that it has not
provided nearly enough three-bedroom units. Before
the war, babies weren't fashionable, and the architects of Fresh Meadows may have been guided in the
allocating of apartment sizes by prewar standards and
expectations. But Stuyvesant Town houses twenty-five
thousand people, and if Fresh Meadows had been built
to the same space standards, almost seventy-five thousand people (God help them!) would be living there.

Fresh Meadows has the further advantage of being
surrounded by public parks. Kissena Park Corridor
touches one corner of the property, and not far away
is Cunningham Park, a playground covering five hundred acres. Within a mile, there are two public golf
courses. To more niggardly real-estate men, these
neighboring facilities would have been an excellent
excuse for reducing the open space in the project, but
the New York Life has set aside six acres studded with
magnificent old trees, mostly oaks, as a park. In the
years to come, the district surrounding Fresh Meadows
will probably be filled with the one- and two-family
houses of speculative real-estate developers, and if it
goes the way of Flushing and Flatbush, it will become
an architectural and social shambles. But a hundred
years from now, Fresh Meadows will, unless it falls
into less conscientious hands, still be as spacious, handsome, and "sweet"—a green island in the midst of

Queens—as St. John's Wood, which long remained an island in the growing welter of London.

Only one public street—188th—goes through Fresh Meadows, and this at the insistence of the Flushing Municipal Engineer's office. The architects have taken some of the curse off this thoroughfare by creating two ovals along it. These are handsomely planted with privet hedge, petunias, and ageratum. In one, white petunias contrast with blue ageratum; the other is a mixture of striking mauves and purples. It is the sort of happy landscaping that recalls Mr. Moses's successful handling of Jones Beach.

Because there was only one through city street to deal with, the planners were not handicapped by having to cope with the standard dreary pattern of blocks, with its wasteful excess of paving and utilities, as were the planners of Sunnyside Gardens, that pioneer neighborhood community developed under the leadership of that enlightened realtor, Alexander Bing, over twenty years ago. The harmony of the residential areas is mainly a natural result of the expanses of lawn, the rows of trees between the houses and the sidewalks, and the islands of park that break up not only 188th Street but some of the cul-de-sac streets, and it is enhanced by the complete absence of the customary rash of private garages. There is ample street-parking space, and there are large parking lots at each of the three shopping centers. In addition, there are four community garages. This contributes enormously to the peace and beauty of the community; where there are individual garages, the land is badly chopped up and the sidewalks are a potential danger for everyone, and especially children, as the cars back across. Here, as elsewhere, the architects took a hint from Sunnyside Gardens. They have often improved upon the hints. In the few instances in which they went

wrong, they followed the unfortunate precedents of
Stuyvesant Town.

Fresh Meadows is perhaps the most positive and ex-
hilarating example of large-scale community planning
in this country. To understand all the things that have
been learned about town planning in the last thirty
years, in both the United States and Europe, you need
only examine the articulation of this design and see
how it provides for the comfort and the aesthetic satisfaction of the inhabitants from the time they wake
up, after a night of peaceful sleep, and put their small
children on the lawn behind the houses—where they
scamper about with reasonable safety on the grass
while the mothers go shopping at the nearest market-
ing center—until everyone goes to bed again. Fresh
Meadows is not just more housing; it is a slice of the
City of Tomorrow—not the futuramic city of Hugh
Ferriss's theatrical (and moonstruck) charcoal architectural sketches but a place that will stand up under
the closest critical inspection. I have looked at it more
than once, and I hope to look again and make a further report.

                                                  1949

TWO

Fresh Meadows, Fresh Plans

One of the things that have bedeviled city planning and housing in New York is the people who are concerned with these arts. They seem to know no middle ground between congestion and sprawl. Working within the city, they usually accept the fact that it is crowded and cluttered, and they think they have accomplished all that can be expected if they leave a little space between their buildings, and plant a little patch of park or asphalt a little strip of playground, without fundamentally alleviating the congestion and the brutal lack of intimacy in the architecture. Working in Westchester or Long Island, they often pretend that each house in a project is a mansion on a big estate, a veritable annex of Arcadia, though the houses may be so close together that you can't look out a side window without seeing your neighbor brush his teeth or quarrel with his wife. In general, the field has been divided between the conscious and unconscious disciples of Le Corbusier, who believes that even a village should be a skyscraper on stilts, and the disciples of Frank Lloyd Wright, who believes that cities should be abolished and that everyone should have at least an acre of land to live on. Neither school is producing the sort of city we all seem to have forgotten, the city in which it is a pleasure and a convenience to live.

Between these two extremes of town planning, there is an aspect of urban building that had hardly been explored before the New York Life Insurance Company decided to make Fresh Meadows, its hundred-and-seventy-acre development in Queens, not merely a bare housing project but a complete community. Not long ago, I said a good many things in praise of the results of this decision, and another visit prompts me to an extension of those remarks. Fresh Meadows is a distinctly urban section, part and parcel of a big city. And because its density of population—seventeen families to the acre—is almost the minimum that can be expected in a city, it sets a basic pattern for the ideal reconstruction of the outer metropolitan area: a vision of harmony, order, and joy. Fresh Meadows is not a temporary refuge from overcrowding; it exemplifies the sort of city planning and building that would decrease the need for rural hideouts and escapist bungalow colonies. If all New York were designed on the same principles, the roads leading out of it would not carry such a weekend load of desperate people looking for a spot of green or a patch of blue or a pool of quiet—the mirage of the great metropolitan desert.

The residential quarters of Fresh Meadows consist entirely of apartment houses. Some are actually groups of single-family two-story units, or duplexes; others are three-story buildings, in which three flats open on each landing; two of them, accommodating over six hundred of the project's three thousand families, are old-fashioned thirteen-story elevator apartments of the type that is now practically standard in Manhattan slum-clearance projects. Though these long, high units are surrounded by a plenitude of open space, they look crowded, and in one vital respect *are* crowded, since this is the only part of Fresh Meadows where children are not allowed to run freely over the ample lawns. The view from the upper floors may be wide, but it is also bleak, and on the north side of the buildings it is a windy one, too. Most of the rooms lack either through or cross ventilation, and there are no open balconies to make up for this lack.

Doubtless, the architects had plausible reasons for the thirteen-story units. They may have wanted to avoid an institutional look, to provide a greater architectural variety than the two- and three- story units alone could do. They may also have wanted to provide a certain number of small apartments, and to satisfy the alleged New York liking for living up high, and they perhaps felt that if they overcrowded the land at this one spot, the rest of the project could be more open. But if what they wanted was variety in the picture, the gigantic palisade formed by the two buildings was not the only way to achieve it, nor the best, for the buildings themselves deny the very quality they seek to establish. I suspect that the real reason for their height is that someone accepted too uncritically the current practice—which has become almost axiomatic in slum-clearance projects—that when elevator service must be provided, thirteen stories is the most economical height. This is the only suggestion that purely mechanical and economic factors, of minor importance, were allowed to take precedence over intelligent and humane design in Fresh Meadows.

Apart from this, the quality of the internal planning and the external design in the main residential quarters is remarkably good. Both the three-story apartment houses and the duplexes are grouped in long rows—the duplexes in straight rows, the others in rows shaped rather like brackets, always placed so that the brackets bend inward from the streets upon which they front. The walls are of red brick; the windows are broad steel casements, somewhat fussily overdivided. The dominant colors—red brick and white trim—are a definite contrast to the white walls and green lettering and awnings that prevail in the project's shopping areas. Since long walls of red brick, even though in-

dented, as they are in Fresh Meadows, and broken by
plantings of ornamental cherry trees, beeches, maples,
and lilacs, can be monotonous, the architects have introduced shallow vertical metal bays, painted white, at
the ends of the three-story units, thus providing an
emphasis that might have been made even more emphatic by finishing some of the lower units in brick
painted white. The design of the three-story apartment houses produces irregularities in the ground plan,
and these irregularities create pleasing contrasts in
light and shade where the walls are at right angles.
The architects have introduced variations in the
straight rows of duplex houses by using different entrance porches—now the little roof is carried on two
slender columns, now it is thrust out from the wall
without support, now it is supported by thin uprights
designed to serve as trellises for morning-glories. Architecturally, these buildings have neither front nor
rear; they are conceived in the round, as honest, three-
dimentional architecture. Flat roofs, incidentally, prevail throughout—a happy contrast to the gaudy gables
in a nearby state-sponsored project.

The street blocks are mainly oblong or square, but
good site planning and an irregular spotting of trees, to
say nothing of curving street corners, take away any
sense of excessive formality or mechanical regimentation, and a number of the units are grouped to form
loose quadrangles, sufficiently open so that the eye can
rove beyond them into the next block. In this way, the
architects have produced both a comfortable feeling of
enclosure and a sensation of freedom and openness.
Thomas Jefferson used the open row in laying out the
University of Virginia, and for a time it was a fashionable stereotype of modern European architecture. Naturally, such a quadrangular grouping pays no attention
to orienting living quarters to give them the maximum
of sunlight, but the feeling of usable space produced

is so superior to that of the rigidly oriented row, and the invitation to the outdoor life so pronounced, that I think the architects have served the greater good. A small block may contain only four building units, a large one may contain as many as nine, but always the open quadrangle is ranged about a wide and tranquil central green, in which there are clumps of trees or a shaded oval asphalted area, forty-five by sixty feet, with benches under the trees, which serves as a play space. Each of these has a small jungle gym and a swing. Each is a convenient, charming, and natural meeting place for mothers as well as children. At the end of a long summer of drought, the grass was still green everywhere. That is proof that the density of population is not too great; when lawns are eroded, it is usually because too many feet have been passing over them.

Sunnyside Gardens, in Long Island City, was the first development to use a combination of apartment houses and row units of different sizes, the first to provide sufficient public open space as an integral part of a project instead of expecting the city to catch up on its heavy arrears in park and play space, the first to demonstrate all over again the greater privacy of row housing as opposed to the confusion and crampedness of the detached or semi-detached house when less than a quarter acre of land is allotted to it. The Fresh Meadows architects thoroughly absorbed and improved upon the lessons taught by the Sunnyside architects—Henry Wright, Clarence Stein, and Frederick L. Ackerman. In their attention to the smallest details, such as the vertical right angle of concrete that conceals the garbage can and the milk bin on the street front of the duplexes, the Fresh Meadows planners admirably underlined the aesthetic lesson of their project—that order, humanely conceived, is the basis of all good design. One of the few things they forgot they are now making up for. Originally, each tenant of the duplexes was provided with only a folded pocket handkerchief of paved terrace in the rear—a plot completely without privacy. Because practically every normal human being, even a hardened New Yorker, is a potential gardener, the management is now permitting the tenants to extend their flower gardens into the lawns, and the resulting brilliant show of colors will increase the loveliness of the green. In some future project, the architects could create an even bigger garden area and plant hedges between the plots, for privacy. This would persuade people to live out-of-doors much more than they do, if indeed they need any persuasion. Sunnyside provided a plentiful allotment of space in which a tenant could garden if he wished. Dean William Wurster recently observed that the private garden seemed to him as essential to good housing as the private bath.

Fresh Meadows was planned after Stuyvesant Town and the neighboring Peter Cooper Village, at a time when the costs of materials and labor were at an alltime peak, and while the work was in hand the costs rose so steeply that there were misgivings about the generous fashion in which the project had been conceived. This may be the reason that in one respect— the size of some of the rooms—Fresh Meadows falls behind the very high space standards of Peter Cooper Village. The smaller living rooms in the duplexes, and some of the combined kitchens and dining rooms, seem cramped, and though the closets are adequate, there is not enough household storage space, for there is neither cellar nor attic space in the duplexes, and trunks and packing cases must be stored in the common rooms provided in neighboring three-story units. If the architects had widened these houses by even two or three feet, the space added would have increased the comfort of the occupants without curtailing too drastically the open area between buildings.

A couple of minor improvements would also have been welcome. One is a horizontal window, or hatch, in the kitchen walls that flank an entrance foyer, so that tiny children could have used this foyer as extra play space and still have been under the eye of their mother while she was cooking. Another would have been putting some of the kitchen-dining spaces on the garden instead of the street side—much pleasanter for the housewife at work, more cheerful at mealtime, and more convenient when one wishes to serve food or drinks outdoors in the summer. A third improvement, and an easy one, would have been to alter the red-and-white color scheme by using a diversity of positive colors in the doorways. Wherever marigolds or morning-glories are in bloom, or a blue deck chair is placed beside a white rear door, the effectiveness of the extra touch of color is noticeable. Unfortunately, flowers and deck chairs belong only to the summertime.

In the three-story apartment houses the architects have, I suspect, built even better than they knew. Twenty years ago, Sunnyside demonstrated that, for a walkup apartment house, a simple, oblong building, without wings and two rooms deep—with two apartments flanking each landing—is the most economical and efficacious type of construction. It wastes no space on corridors or overlarge foyers, and it provides through ventilation, or cross ventilation, for every room. In the intervening years, that demonstration has been generally ignored. New York architects have tied themselves into knots trying to increase the number of apartments they can carve out around a single elevator shaft, but the more elaborate and ingenious these plans, the worse the results. The Fresh Meadows three-story buildings, in which three apartments open off each landing, are the only satisfactory variants I

have seen on the Sunnyside formula. They provide
three- and four-room apartments of better shape and
slightly greater size than those in the thirteen-story
Fresh Meadows units, and there is not a square foot of
waste space anywhere in the buildings. The quality of
these units, whose upper floors are reached by stairs
more gradual than is usual, is proved by the fact that
instead of renting more cheaply than the ground floor,
the upper floors rent readily for the same price, and
will probably continue to even when the housing shortage abates.

There is no reason the concrete-frame construction
and floor plan of the three-story houses should not
be used for elevator apartments six or seven stories
high, possibly at a lower cost per room than in the thirteen-story buildings and affording the tenants far
greater comfort and beauty of outlook. The spotting
of six or eight such buildings around the project in
place of the two tall ones would have added to the variety and interest of the scene; the density of seventeen
families an acre could have been achieved without a
major encroachment upon the generous open spaces
provided by the plan that was adopted. If the architects had thus carried their demonstration this one
step farther, they could have shown their confreres
that tall buildings are not necessary even in urban housing.

But all in all, the verdict must be "Well done!"
The residential section of Fresh Meadows has some
of the peace and order, derived from the absence of
all the things one doesn't want and from the presence
of the things everyone does need, that Henry James
once so beautifully described: a place composed of
buildings "all beautified with omissions," a place
where "all sorts of freedoms" are enjoyed, a place
where one may have one's fill of sunlight and green
lawn and sky and trees, where there is aesthetic and

spiritual composure. As individual buildings, these duplexes and apartment houses are not superlatively good architecture; indeed, their aesthetic components are fairly undistinguished. But the whole that they form is better than the sum of its parts. Skillfully put together, these form not just "housing" but a beautiful community—complex and many-sided and serene. This is the important difference between Fresh Meadows and all ordinary housing projects.

1949

THREE

U N Model and Model U N

At the end of the preliminary report to the General
Assembly of the United Nations on the subject of that
organization's permanent headquarters, the Board of
Design Consultants says, "It is with some trepidation
that these plans are submitted for the consideration of
the General Assembly. They are an abrupt crystalliza-
tion in the course of the creative process of continuous
experiment." In dealing with plans put forward in this
disarming manner, a certain degree of humility, not to
say fear and trembling, should probably modify the
judgments of the critic. Certainly it is with complete
sympathy for those compelled to show their work in
an unfinished state to an aggregation of officials who,
in the nature of things, can have only an imperfect
understanding of the problems involved that I undertake to examine this preliminary survey. Except for
a couple of soft renderings of the finished structures in
the dim Hugh Ferriss style that was a characteristic of
the 'twenties, the report is an extremely competent,
workmanlike job.

When one considers the problems the architectural
team has been up against, one wonders, to begin with,
at the speed and adroitness with which they have gone
as far as they have. The Board of Design Consultants,
whose chairman is Mr. Wallace Harrison, is made up
of nine architects and one engineer. This group is as-
sisted by seven other architects, a landscape architect,

and eight other engineers. Fifteen countries are represented on this panel, and not all the architects speak
English, either. Getting even a Tower of Babel out of
such a group would have been an achievement. Inevitably, to accomplish all that has been done required a
certain concentration of responsibility, and, almost
as inevitably, at this preliminary stage, one sees the
preponderating influence of—I was going to say Mr.
Harrison, but it would be more accurate to say New
York. If more time had been allowed the Board, other
influences would undoubtedly have had a better
chance to make themselves felt; there is nothing more
time-consuming than genuine cooperation.

The Board of Design Consultants did not begin to
function until February of this year. "Speed," we read,
early in the document, "was the essence of the problem." "Some fifty basic designs were created, criticized,
analyzed, and resynthesized," we are told a little later.
Those two statements, taken together, account for the
fact that when it became necessary to reach a decision,
the Board fell back upon the architectural stereotypes
of the early nineteen-thirties. So far, their headquarters is a combination of Le Corbusier's breezy City of
the Future and the businesslike congestion of Rockefeller Center, a blending of the grandiose and the obvious. The present designs for the buildings will no
doubt be improved, perhaps radically modified. But
the problem created by the site granted the United Nations is much harder to correct. Whether there is wisdom enough to overcome these handicaps, I do not
know.

As I have just pointed out, the worst problem facing
the architects who are laboring on the United Nations
project is the site. Architects should not, of course, be
blamed for the errors of their clients. Is this site the
right one for such an institution, and is the area adequate? There is no question in my mind but that the

headquarters for the United Nations should be in the heart of a metropolis. There is also no question but that an organism that is bound to grow and to attract to its neighborhood other institutions fostering international activities should be a real city within a city, dominating its site even more conspicuously than Vatican City does. The site for such a headquarters should eventually embrace around a thousand acres, or at least be not less than the size of Central Park, which is around eight hundred and fifty acres. (The United Nations site, in its present form, is only seventeen and a half acres.)

Business organizations dare not contemplate such huge schemes, and few national governments would dare to, either. But this is a world organization, and if it is not to fail us, it must grow rapidly both in collective wisdom and in authority, and that authority should be visible. In a city of skyscrapers, only a megalomaniac would demand a higher group of buildings than those on the rest of the sky line; to set itself sufficiently apart from the welter of other buildings—to say nothing of ensuring room for growth—the United Nations headquarters needs plenty of space.

Don't think that an ample and advantageous site could not have been found on Manhattan Island. What with the prewar exodus of our residential population and the rotting of slum and industrial properties, one could have picked out two or three magnificent sites. Perhaps the best, and the ripest (that is, the rottennest), would have been the area immediately south of Washington Square, a region almost destitute of tall buildings, let alone substantial ones. Imagine an assembly hall on the axis of Fifth Avenue. It would fill the eye, and if there were no skyscrapers near at hand to compete with it, it would look thoroughly important, too; indeed, no other big building in the city would be set off to better advantage. Obviously, the

United Nations would have needed the city's and the federal government's help to acquire such a site, but in the course of a generation the ten or fifteen acres it would have started with could have been added to without greatly disrupting the life of the city. There should have been plenty of time to ponder what form the site and buildings of a permanent headquarters should take. But the United Nations jumped at the first fleabite of land that was offered it hereabouts, and that is unfortunate, since it could be taken as an implication of a sense of uncertainty, of impermanence.

Now let us take a close look at this fleabite. It is in the old slaughterhouse district on the eastern edge of the island, and it covers the area between Forty-second Street and Forty-eighth Street, bounded on the west by First Avenue. Except for the Housing Authority's new building, on Forty-second Street, and the nearby skyscraper apartment hotels in Tudor City, the whole neighborhood is a dismally blighted area. Flanking the site on the south, Tudor City rises like a ruddy Maxfield Parrish dream above Forty-second Street, accentuating the steep, rocky slope of the land down to the river level that is one of the difficulties of the site. The apartment buildings and the slope both say, "Keep Off." The East River Tunnel ventilator building and the immense power station at Fortieth Street would make expansion of the site in that direction impossible. To the north, at Forty-eighth Street, an abrupt rise in the terrain, topped by the tall apartment buildings of Beekman Place, blocks growth in that direction.

The United Nations has, happily, permission to build east over the East River Drive as far as the bulkhead line, but the only direction it can really expand is westward. The valuation of the land here being what it is, this would be an expensive proposition, but one can still hope that this expansion will take place. The city has agreed to spend seven and a half million dollars, which is almost as much as the site cost, simply to drive a tunnel beneath First Avenue from Forty-first Street to Forty-eighth, so that through traffic will be kept away from the western border of the site. The city has also undertaken to provide a strip of park from Second Avenue to First along Forty-seventh Street to make what someone has quaintly and irrelevantly described as an "approach" to the head-quarters. None of this really makes sense unless the city is prepared to go much further. The present approach to the United Nations is a sordid slum, and it cannot be left to the erratic beneficences of free enterprise to provide the only approach that would be aesthetically sound; namely, a group of related buildings, conceived in accordance with the function and purpose of the United Nations headquarters itself. To think that a strip of formal park will create the necessary atmosphere is not to think at all.

Before Mr. Rockefeller bought the United Nations site, the land had been assembled by a group of private operators, who had engaged Mr. Harrison to prepare plans to develop it as an apartment-house and business center. The rise in the cost of building may well have made this project, conceived during the war, seem a dubious prospect, and the fact that Mr. Harrison was one of the major architects for Rockefeller Center may have made it a little easier for the principals in the present undertaking to get together. Once Mr. Rockefeller's gift was accepted, the appointment of Mr. Harrison—about whose outstanding abilities as architect and organizer there is no question—as the chief architect of this enterprise seemed an almost inevitable step. This essay in genealogy is not designed to look a gift horse in the mouth. It is merely intended to show that this particular site, though it might have served business well, is too cramped adequately to serve the United Nations even as working quarters. And what can be built on it will hardly be a fitting symbol of a just and orderly world. What might be a good short-term realty speculation is not necessarily a good long-term investment in the interests of the welfare and the comity of the peoples of the earth. In another article, I hope to have something to say about the absence of symbolism in the architecture. There is, however, plenty of unconscious symbolism in what has happened so far. Because Mr. Bernard Baruch was once a Wall Street man, it was probably bad symbolism to let him head up our Atomic Energy Delegation to the United Nations, even though he was daring enough to support a plan for socializing atomic energy on a world scale. And it was bad symbolism to let Mr. Rockefeller get mixed up with the United Nations headquarters. Mr. Rockefeller is a benevolent philanthropist who at one moment restores forgotten Colonial capitals and at another presents the City of New York with a handsome public park. But to some of our more difficult brothers overseas, Mr. Rockefeller is Monopoly Capitalism, and the fact that it was he, and not the City of New York or the federal government, who gave this site to the United Nations will not, unfortunately, lessen their suspicions and animosities. If, when the project is finished, the United Nations headquarters should, by some fatal chance, look like Rockefeller Center, that would produce both a mischievous and a misleading impression.

Briefly, then, I should say that the architects have been afflicted with a bad site, and that unless the government, local or national, helps out on a very large and costly scale, they cannot do much about improving it. They are not to blame for the unsatisfactory

plans they have submitted, except in one respect:
the report of these honorable and intelligent professional men should have advised their client, the United
Nations, that in using this site it was doing something
it would probably bitterly regret.

                                                              1947

FOUR

Buildings as Symbols

The model of the proposed United Nations headquarters probably doesn't do justice to the more positive virtues of the architecture, such as the gleam of
the glass in which the buildings will be partially
sheathed, but it gives a rough idea of the dimensions
and relationships of the proposed buildings and open
spaces. Since, however, my comments on these buildings will touch on their inner functioning as well as
their outward order, a better reference work for this
treatise on symbolism in design might be the official
preliminary report on "The Permanent Headquarters
of the United Nations" (United Nations Publica-
tions: Sales No. 1947.1.10). The problem that faced
the distinguished Board of Design Consultants, of
which Mr. Wallace K. Harrison is the head, was determined by two main considerations: the size and nature
of the site, and the work that the United Nations is doing now and may be expected to do in the future.
Recently I made a few regretful remarks on this site.
The architects, to judge by the report, are better sat-
isfied with the area than I am, for they say, "The East
River site, extending fifteen hundred feet from Forty-
second Street to Forty-eighth Street, and from First
Avenue to the edge of the water, has sufficient scale
for applying the fundamental elements of modern
urbanism—sunlight, space, and verdure. Protected
by the wide expanse of the East River, the site has

breadth enough to be made into a living unity of strength, dignity, and harmony."

Those are brave words, but they unfortunately recall the observations of our first city-planning commissioners when, in 1811, they apologized for the lack of parks in their plan for Manhattan and suggested that the "large arms of the sea" would provide all the open space necessary. The Board of Design Consultants are, I am afraid, a little like Browning's Last Duchess, "too soon made glad, too easily impressed." One could readily forgive them such amiable faults if they had succeeded in making the best possible use of the seventeen and a half acres that have been put at their disposal. Severe limitations of the sort encountered on this inadequate site are sometimes spurs to the imagination, so I examined the model and the plans carefully to see whether in this case they had had any effect. With great reluctance—indeed, with pain and embarrassment—I must cast a negative vote. Symbolically, these buildings are far from being an admirable expression of the idea of the United Nations, and functionally, they do not make the necessary provisions for extension, change of purpose, and future development.

The architects' preliminary scheme provides for five structures. At the north end of the site will be a large office building that will house the delegations. Being on an east-west axis, it will form a sort of wall flanking the group of apartment houses to the north, on Beekman Place, and it will be completely set apart from the other structures by a big park. Toward the south end of the site will be a group of four buildings. The dominant one will be that of the Secretariat, a rectangular slab of skyscraper forty stories high, halfway between the river and First Avenue, and running from a point a bit north of Forty-second Street to a point halfway between Forty-third and Forty-fourth

Streets. Directly north of this, at about the center of
the site, will be the slightly fan-shaped General Assembly Hall—a great auditorium, whose axis will be parallel to First Avenue, with an entrance for the general
public on the north, reached by a roadway through the
park from First Avenue. This building will be broad
and, because of its function, nowhere near as tall as
the two others I have just mentioned. The buildings
of the Secretariat and the Assembly Hall will be connected by a wide unit, even lower than the auditorium,
lying to the east and to be devoted to the councils and
committees of the United Nations. South of all this, on
Forty-second Street, near First Avenue, another structure is already in existence—a small office building
erected by the New York Housing Authority and
turned over to the United Nations before it was finished. Plainly, this last is a mixed blessing, very useful
at the moment, when the United Nations needs administrative quarters in the city, but an obstacle to any
free development of this part of the site.

In trying to view this design as a whole, one quickly
becomes conscious that, aesthetically speaking, a
whole does not exist. Instead, there are five visually
unrelated members, one of which, the skyscraper for
the Secretariat, is so tall that the lesser ones could not
have any impact on the spectator. (This, by the way,
is the only unit for which the architects have provided
a visual approach from the west; its entrance is plump
on the axis of Forty-third Street.) Despite the variety
of their national backgrounds, the architects have apparently not been able to shake off the stereotypes of
New York architecture. There are, of course, cases in
which a tall building, towering above the flat prairie,
like Goodhue's State Capitol in Lincoln, Nebraska,
can serve as a definite architectural symbol. But to
rely upon creating a similar effect in a city of skyscrapers is either to miscalculate the visual relation-

ships completely or to scorn them altogether. The
neighboring R.C.A. Building is seventy stories high,
the Chrysler Tower is even higher, and the nearer-at-
hand pinnacles of Tudor City, set just southeast of the
United Nations site, on a bluff that rises above it,
must diminish the effect even of the forty stories of the
Secretariat Building.

In a city whose office buildings, apartment houses,
and hospitals are on a colossal scale, the only way to
set buildings off and to give them dignity and value is
to surround them with an ample area of trees and gardens and to keep the buildings themselves reasonably
low, so maintaining the human scale, which the other
parts of town sadly lack. Even the fountains and the
tiny strip of flower bed that leads from Fifth Avenue
down to the plaza at Rockefeller Center produce an
aesthetic effect out of all proportion to their size, precisely because of the modest height of the buildings on
both sides. In the United Nations headquarters, there
is not only no human scale, there is no transition from
the intimate to the monumental. And it was a mistake
to make the Secretariat Building the monumental,
dominant structure instead of the General Assembly
Hall and the Conference Building, which should be
the focus of visual interest as well as the symbol of
political authority. If the Secretariat Building will have
anything to say as a symbol, it will be, I fear, that the
managerial revolution has taken place and that bu-
reaucracy rules the world. I am sorry that the architects have apparently taken Mr. James Burnham's
discouraging thesis as an axiom, for the United Nations is an attempt to make other ideas prevail.

Doubtless the Beaux-Arts architects of the last generation would have tried to make the United Nations
headquarters look richly palatial and ancient—a
classic forum or a baroque plaza—and the result
would have been as grisly as the new government

office buildings between Pennsylvania Avenue and the
Mall in Washington. Perhaps we should simply be
thankful that Mr. Harrison and his colleagues have
saved us from that fate. But the Beaux-Arts architects
would at least have tried to do one thing right; they
would have taken care that the important buildings of
the group would be visible at a distance, would have a
handsome setting, and would make a powerful aesthetic impression. The United Nations headquarters
should not look like a group of temples and basilicas,
but it likewise should not look like a few forlorn and
temporary "taxpayers" nestling under a group of office
buildings till the rest of the site is ready to be covered by skyscrapers. That is just another kind of imitative formalism. In view of the difficult conditions
under which they have worked, one could forgive Mr.
Harrison and his associates for their mistakes of detail.
Their real failure is of a different order; they have
failed to create a fresh symbol, and that failure is more
serious. Their new World City is just a chip off the old
block. Even the architectural huggermugger of the
World's Fair of 1939 produced a Trylon and a Peri-
sphere, those slightly embarrassing efforts toward symbolism, and if Mr. Harrison, their designer, is trying to
live them down, all I can say is that he has swung too
far in the opposite direction.

Except for the series of esplanades, ramps, and terraces that sweep upward from the landing stage at the
river's edge to the site itself, on top of the bluff that
fronts the water, the disposition of the buildings and
open space can in no respect be called satisfactory.
When Mr. Harrison handed in the preliminary plans to
the Advisory Committee, of which Mr. Warren R.
Austin is chairman, he presented them with these
words: "The world hopes for a symbol of peace; we
have given them a workshop of peace"—almost as if
to say that the symbolic part of architecture, the im-

pression it makes on the mind and the senses, is of
minor importance. If he did not mean to say this, the
architecture has said it for him, for certainly two tall,
unrelated office buildings, an auditorium, an all but
invisible conference building, and a small administration building are hardly an impressive group in this
city of towers. Considering the genteel classicism of
the winning design in the competition for the Palace of
the League of Nations, the modernity of the United
Nations plans must have seemed to many of the architects involved—especially to Le Corbusier, whose austere design for the Palace was rejected by the Geneva
jury—a great triumph. Time, however, changes
everything, including the meaning of "modern." To
say that a group of buildings is not fake Greek, bastard
Roman, cast-iron Gothic, trite Renaissance, or simply
warmed-over hash is no longer sufficient commenda-
tion. By now, these negative virtues, along with sunlight, air, and verdure, must be taken for granted. This
should be our point of departure, not our goal.

And what about practical considerations? The Secretariat Building is designed to hold forty-four hundred
workers; the number that must be housed now is only
twenty-five hundred. An obvious alternative to the
oversized skyscraper, which would dwarf all the other
structures on the site, would be to build two ten- or
twelve-story structures and to add from two to four
more as the office space is required. These buildings
would provide a modest background for the General
Assembly Hall and the Conference Building, which,
as I have said, should be the central features of the
scheme. Such a decentralization of the Secretariat's
offices would be practical as well as architecturally and
symbolically satisfying. For one thing, much less interior space would have to be wasted on elevator shafts;
for another, less time and expense would be frittered

away in vertical transportation, the costliest form of urban locomotion; finally, it would disperse, rather than concentrate, pedestrian and motor traffic during the peak hours. If the architects are bent on imitating Rockefeller Center, they should copy not its bad features but its progressive ones, especially the scale and proportions of the Eastern Airlines Building.

There are, I should emphasize, some thoroughly commendable points in the present design of the Assembly Hall and the Conference Building. The latter, for example, with its wide windows looking out on the river, with its top-level restaurants and sunny terraces, will offer the delegates the recreative resources of light and space and wide horizons—resources that will serve them, in their more harried moments, better than aspirin, whiskey, or sleeping capsules. I do not, however, understand why the designers have not provided the delegates with a private waterfront promenade, where they might stretch their legs, along with little bosky recesses, where they could, in solitude, conduct private negotiations or just recover their sense of humor.

Can it be said that the functional, or "workshop," merits of this headquarters are so extraordinary that they can make one overlook its aesthetic shortcomings? The more I inspect the plans, the more doubtful I am. Take the vital matter of site planning. To make the fullest use of the site, all wheeled traffic should be put underground, especially since ample provisions have been made for parking vehicles there. By doing that, the architects would have avoided squandering no small part of their precious ground area on vehicular driveways and turn-arounds. These will interfere with all the pedestrian approaches from the First Avenue frontage, which is, after all, the natural point of approach to the site. As I pointed out in my first article on the United Nations project, a tunnel under the portion of First Avenue alongside the area will siphon off through traffic, and the surface level will accommodate only traffic that concerns the United Nations. But were the United Nations traffic to be put underground too, certainly half the width of First Avenue, granting the permission of the city, could be added to the area that will be converted into a park. If there is one axiom in urban planning that no modern scheme should ignore, it is that there should be a separation of legs and wheels, of walkers and drivers. If the planners haven't yet come abreast of the plan for the New Jersey town of Radburn (1929), they might have learned a lesson from Central Park (1857) or, for that matter, from Leonardo's plans for Milan (circa 1500).

Take still another aspect of the matter, one even stranger for a group of architects professedly wedded to functionalism. The plans call for a library with space for a million and a half books. (I understand that after the architects' report was issued, a cut in their budget impelled them to revise downward their plans for this.) What reason is there to think that an institution of world importance will not, in a generation or so, want space for double the originally projected number of books, or that it won't attract scholars from all over the world, who will need far more seating capacity than is now provided? Eventually, it would seem, the library will have to limit its acquisitions or move. Neither the unsatisfactory Widener Library, at Harvard, nor the Columbia Library is less flexible in this respect; indeed, compared to this scheme, Carrere & Hastings's library, that classic monument at Fifth Avenue and Forty-second Street, is a masterpiece of functionalism. The report says that room must be provided for expansion, but I can find no evidence in the plans that such provisions have been made. Here, as elsewhere, there is evidence of the failure to respect fundamental principles. The report uses all the right words to describe the architecture; it emphasizes integration, organism, flexibility, expansion, and so on. But in the plans themselves one too often sees just the opposite—rigidity, confinement, lack of sound provision for growth.

Some of the weaknesses in the project will doubtless be corrected before the plans are passed on to the contractors. Knowing the immense ability of Mr. Harrison and his fellow-consultants, I do not doubt that the finished structures, even if they are not radically modified, will have many excellences that no model, plan, or diagram can indicate. Whether the basic errors will be corrected is another matter. And that is supremely important. For whatever the United Nations headquarters should do or say, however many people or functions it must serve, the buildings should proclaim with a single voice that a new world order, dedicated to peace and justice, is rising on this site. These buildings should be as beloved a symbol as the Statue of Liberty, as powerful a spectacle as St. Peter's in Rome. Such symbols cannot be created by falling back on clichés, like statues, domes, and skyscraper towers, and they cannot be conceived overnight. Short of flatly rejecting the site, the architects should have worked out the right scale for their present project, suggested an effective way of taking care of future requirements, and then set their most imaginative members to work on the problem of symbolism, which is, at bottom, the problem of public relations for the new world order. Better the right thing a little late than the wrong thing on time.

1947

FIVE

Magic with Mirrors

The dominant building in the United Nations group
has been visible in its glassy glory for many months.
Though the Secretariat Building is not yet finished or
fully equipped, it is occupied and in operation. If any
of the staff should break a leg or feel the pangs of
childbirth, the well-organized hospital on the third
floor could probably set the leg or deliver the baby
without the patient's having to leave the premises.
Now that there's a cafeteria, a bank, and a post office
on the premises, too, the building provides for most of
the daily non-official needs of its occupants. As a work
of art, though, the Secretariat Building is a teaser.
From people's opinions of its architectural significance,
one can make a fair estimate of their aesthetic sophis-
tication, their human insight, their social values, even
their moral standards. A Southern editor, defending an
archaic and unfunctional hospital his community is
erecting, observed that it was not, thank heaven, as
ugly as the United Nations Secretariat. Yet architects
as able as Richard Neutra and William Wurster have
pronounced the building a great achievement. Since
there is no point in hiding one's opinion out of respect
for recognizable and presumably living people, I will
say that the Secretariat Building seems to me a superficial aesthetic triumph and an architectural failure.
A few more triumphs of this nature, and this particular school of modern design might be on the rocks.

In this building, the movement that took shape in the mind of Le Corbusier in the early nineteen-twenties—and that sought to identify the vast and varied contents of modern architecture with its own arid mannerism—has reached a climax of formal purity and functional inadequacy. Whereas modern architecture began with the true precept that form follows function, and that an organic form must respect every human function, this new office building is based on the theory that even if no symbolic purpose is served, function should be sacrificed to form. This is a new kind of academicism, successful largely because its clichés readily lend themselves to imitation and reproduction. In the present instance, it has brought into existence not a work of three-dimensional architecture but a Christmas package wrapped in cellophane. Functionally, this building is an old-fashioned engine covered by a streamlined hood much embellished with chromium. The package has been conceived with what would appear to be not even a motorcar stylist's interest in the contents.

From a distance, the Secretariat Building, two hundred and eighty-seven feet long, seventy-two feet wide, and thirty-nine stories high, is a great oblong prism of glass, marble, and aluminum. It connects, on its lower levels, with the General Assembly Building, to the north of it, and with the Conference Building, to the east and almost invisible except from the river. But by reason of its bulk and height, this huge slab is visually detached from them and reduces them to insignificance. The smallest buildings in Rockefeller Center are far enough away from the enormous R.C.A. Building and are sufficiently supported by buildings of intermediate height not to seem runty, but there are no such spatial gradations between the midgets and the giant in the United Nations composition; the success of the whole group depends almost solely upon this central building. The exterior of the Secretariat is much less complicated than that of the R.C.A. Building, for there are no recessions or setbacks. At the north and south ends, this prism is a smooth, windowless sheath of mottled white marble; on the east and west faces, it is a smooth wall of green glass framed in aluminum. Even the spandrels and frames of the windows do not break the surface; in fact, the only interruptions in it are four horizontal grilles, each one a story high and running the full width of the facade. These grilles, set at intervals, conceal the several installations of elevator and ventilating machinery. The lattice effect they create is repeated above the roof, to a height of over twenty feet, to conceal the penthouse, which also contains machinery. This manner of visually dividing the building was the object of adverse criticism in a recent architectural symposium. But on the whole the change of form seems to me a happy way of externally acknowledging a change of interior function, and even the latticework at the top, though costly, is a justifiable liberty in a rigidly restricted design.

In one sense, the Secretariat is the fulfillment of a long-cherished dream. Ever since Sir Joseph Paxton built the Crystal Palace in London, precisely a hundred years ago, the idea of continuing that development in steel and glass has haunted people's minds. When, in 1898, Sir Ebenezer Howard outlined his garden city, he thought that its whole shopping district might be an extended Crystal Palace, forgetting the tropical temperatures that even an English summer generates in a glass hothouse. In 1921, Mies van der Rohe developed plans for a skyscraper conceived in all innocence in steel and glass, with glass walls from floor to ceiling, without (as far as could be seen) pipes or utility ducts or any protection against damage by fire to the steel beams and columns, without any spandrels to conceal these utilitarian elements or any device to lessen the complete feeling of insecurity as one approached the outer walls.

Unfortunately, glass and steel are not wholly satisfactory building materials. If steel is not insulated from heat, it expands and contracts in a fashion that presents serious problems, particularly in a tall building placed where other buildings or trees do not modify the climate. Glass transmits not only light but heat, and unless windows are completely sealed, they admit air in a high wind. And, as last November's hurricane once more demonstrated, particularly in the case of the Secretariat Building, large sheets of glass are perilously breakable. Therefore, the development of the skyscraper became possible only when architects learned to give as much attention to heat-and-fire-resistant materials as to the revealing qualities of glass and the structural possibilities of steel. Glass and metal do not burn, but they crack and buckle in the heat of a fire; the Crystal Palace was demolished by flames in the nineteen-thirties. But the massive masonry of the ancient stone-and-glass cathedrals of Europe stood up under both fire and bomb blast during World War II while the buildings around them were reduced to cinders and rubble. Glass not only admits a great deal of heat on sunny days, even when the windows are closed; it likewise radiates heat to the outer air on cold days. In his own design for the Secretariat, Le Corbusier proposed to overcome these defects with two special devices. One was the permanent sun screen, or *brise-soleil,* which was used on the sunny sides of the Ministry of Education and Health Building in Rio de Janeiro in 1937—a building on which Le Corbusier served as consultant. The other was a double glass wall, inside which he intended to circulate cool air in the summer and hot air in the winter. When these elements in his design were thrown out, he wrote an in-

dignant letter to Ambassador Warren Austin, the head
of the United Nations building committee, declaring
that the steel-and-glass building that has now been
erected would be uninhabitable. There was good reason to avoid the *brise-soleil* in New York's climate,
since menacing icicles might form on it, and there is
also reason to avoid a double glass wall anywhere,
since the cost of cleaning what would amount to a continuous double window would be enormous. But there
was an even better reason for turning down this wall.
Glass should give one a clear view of the outside
world, and in the form of the window—that admirable invention—it provides direct contact with fresh
air and sunlight. A solid glass wall sacrifices one of
these advantages, and a double glass wall sacrifices
both.

But even without Le Corbusier's devices the Secretariat Building is quite different from the all-glass
structure, unashamedly revealing its interior, that Paxton's Crystal Palace was and van der Rohe's skyscraper hoped to be. Because the glass is green (the
color is supposed to lessen the transmission of heat),
the east and the west sides of the structure, viewed
from the street, look dark and opaque, not light and
transparent. So, aesthetically speaking, the main function of these great glass walls is to serve as a mirror in
which the buildings of the city are reflected, in which
the western sky sometimes plays in delicate counterpoint to the eastern sky. No building in the city is
more responsive to the constant play of light and
shadow in the world beyond it; none varies more
subtly with the time of day and the way the light
strikes, now emphasizing the vertical metal window
bars, now emphasizing the dark green of the spandrels
and underlining the horizontality of the composition.
No one had ever conceived of building a mirror on this
scale before, and perhaps no one guessed what an

endless series of pictures that mirror would reveal. The aesthetic effect is incomparable, but, unfortunately, when the building is most effective as a looking glass it is least notable as a work of architecture.

The architects, probably not realizing that their building would become a mirror, clung tenaciously to the two-dimensional quality of the exterior. Except for the grilles, there is no hint of a third dimension in those sheer, unvarying walls. Yet when the day is dark or when night falls, the building gets the better of the designers' intentions, for the stabs of light on the ceilings of the offices add an unexpected liveliness to the west façade, making it almost the equivalent of a starry sky. That same liveliness might have been at least suggested by day if the architects had used white Venetian blinds to introduce a minor variation in the pattern of the façade. Instead, they chose gray blinds, which are considerably less efficient at repelling heat, and when the blinds are lowered, the wall of glass retains its impenetrable two-dimensional quality.

Here, then, is the Secretariat Building from the outside: two thin white vertical marble slabs, connected by two vast glass mirrors that are broken only by horizontal white aluminum grilles; a building chaste, startling, fairylike in its cold austerity, a Snow Queen's palace, exhaling by night a green moonlight splendor. Paraded as pure engineering and applied geometry, this new skyscraper proves really to be a triumph of irrelevant romanticism. If anything deserves to be called picture-book architecture, this is it, for all the fundamental qualities of architecture seem to have been sacrificed to the external picture, or, rather, to the more ephemeral passing image reflected on its surface. Should one look behind this magician's mirror, one should not be surprised to find, if not a complete void, something less than good working quarters for a great world organization.

In planning the Secretariat, the architects were not, like most designers, confined to the constricted space of the conventional Manhattan lot, or even block, and they could thus have designed a freestanding building, or a series of freestanding buildings. There are not many freestanding buildings in America, and most of them are tall and highly uneconomic towers, despite the excellent precedent set long ago by the oblong Monadnock Building, in Chicago, that last masterpiece of masonry by Burnham & Root. The architects of the slab-sided Secretariat have ignored this precedent by putting a thirty-nine-story skyscraper on their ample plot, in the manner of a real-estate speculator trying to get the maximum possible amount of rentable space. A freestanding building can have light on all four sides, but the architects of the Secretariat have blanked out two sides of it with solid walls of marble.

A while ago, Mr. Wallace Harrison, who was the chairman of the United Nations' Board of Design Consultants, explained the decision to build a single tall skyscraper in an address to his brother architects at the Royal Institute of British Architects. "We have found by experience," he told them, "that with conditions similar to those on Manhattan Island a building twenty-five to forty-five stories high is the most efficient and economical." Obviously, the conditions he referred to are the conditions of corporate financing and realty speculation, the conditions that make possible a handsome profit. The United Nations is a nonprofit institution. Mr. Harrison then gave the show away by remarking that when one considers only mechanical requirements, "every skyscraper must be built in units not more than fifteen stories high. . . . Thus at approximately every fifteenth floor you have a 'basement' for water tanks, elevator and air-conditioning machinery, and fire protection." In other words, the most efficient and economical height for an office building is not twenty-five to forty-five stories but fifteen. Mr. Harrison and his colleagues have proved the point by inserting those grillwork "basements" only eight stories apart in the Secretariat Building. As for efficiency and economy, he has forgotten that the taller a building is, the greater the space that must be wasted in elevator shafts bypassing the lower floors. Had the several units that are now stacked one upon another to produce the Secretariat been separate buildings, they not merely would have achieved the economy of which Mr. Harrison made so much but would have scaled the Secretariat unit down sufficiently to give visual emphasis to the General Assembly Building (which is to be finished in 1952) and to arrive at a more unified design for the buildings and open spaces. Such a system of related low buildings would also have provided an orderly means of future expansion for the Secretariat, which a single oversize building does not.

Apparently, though, the Board of Design Consultants were hypnotized by Le Corbusier, and Le Corbusier has long been hypnotized by the notion that the skyscraper is a symbol of the modern age. But the fact is that both the skyscraper and Le Corbusier are outmoded. Skyscrapers conceived without respect for human scale or insight into human requirements and values are indeed symbolic, but they are symbols of the way specious considerations of fashion, profit, prestige, abstract aesthetic form—in a word, "the package" of commerce—have taken precedence over the need of human beings for good working and living quarters.

What we have, then, is not a building expressive of the purposes of the United Nations but an extremely fragile aesthetic achievement, whose main lines conform to the ideals of a boom period of shaky finance and large-scale speculation. This sort of modernism goes only skin deep. As a conscious symbol, the Secretariat adds up to zero; as an unconscious one, it is a negative quantity, since it symbolizes the worst practices of New York, not the best hopes of the United Nations. So much for the outside of the building—impeccable but irrelevant. The inside of this package does not even live up to the elegant wrappings. On that matter, I shall presently have more to say.

1951

SIX

A Disoriented Symbol

Viewed from without, the thirty-nine-story United Nations Secretariat Building, whose east and west fronts form oblong green mirrors, is the glass of present-day fashion. But, once inside it, one discovers that it can make no claim whatever to being the future mold of form. So far from being the model office building it might have been, a light and a lesson to all builders, it is really a very conventional job, and the mistakes in conceiving and planning have unfortunate results upon the comfort and routine of the people who work there. To paraphrase General Bradley, this is the wrong sort of building, in the wrong place, facing in the wrong direction, for the wrong purposes.

The interior has, however, certain striking and even ingratiating features. One arrives at the main lobby, reached from the west, by a semicircular driveway or the walk that parallels it. Though full-grown trees have been planted about the plaza in which these approaches are set, the only object that actually helps to give the structure scale is the mediocre little left-over building to the southwest, built for the New York City Housing Authority and now the United Nations library. By placing the tall and narrow Secretariat parallel to the East River, the architects have included in the vista for all who approach the building from the west the jeering, jangling hulk of the plant that ventilates the Queens Midtown Tunnel and the towering chimneys of the Edison works, farther to the south. (From the entrance of the General Assembly Building, under construction immediately to the north, this industrial area will be even more prominent.) As one passes a side entrance, one gets a pleasing glimpse of the East River through windows that reach from floor to ceiling on the east and west sides of the building. This entrance is, by the way, the logical place for the visitor to enter the building, for the Information Desk has been placed here. The main entrance is marked on the outside by a curving marble bay and an irregularly curved aluminum marquee that, alas, re-called the *"modernique"* of the Paris Exposition of 1925. The absence of a fine sense of scale here is conspicuous, for the entrance, only one story high, is crushed to insignificance.

On the inside, the ceiling, low enough as it is, seems much lower because of the length of the lobby. The interior walls and columns parallel to the east and west green glass façades are faced with green marble. Those parallel to the north and south façades, which are unbroken surfaces of mottled white marble, are faced with the white marble. The floor, which is made up of large squares of black and white terrazzo, calls attention to itself too emphatically. The happiest feature of the lobby, humanly speaking, is the long bench at the south end. This should be a convenient place—like the information desk at Grand Central—for meeting people. The lack of such a feature is a weakness in most office-building design. (Some day, perhaps, this bench will be turned to face the splen-did East River view.) Otherwise, the interior is in the frigid, impeccable taste that is the modern equiva-lent of what the architectural bureaucrats in Washing-ton seek when they invoke the classic. If you like the Mellon Gallery interior, you should like the Secreta-riat interior, even if its proportions are not so noble.

The lobby, the first of five floors devoted partly or wholly to the social functions of the building, is connected by escalators and elevators with an underground garage and passages—one of which leads to Forty-second Street—as well as with the bank, the post office, the health clinic, the press center, and the temporary cafeteria. (As soon as the Conference Building is finished, the cafeteria will be shifted to its summit.) At the top of the Secretariat is the Secretary-General's floor, which is carpeted and includes a small but commodious apartment for the Secretary-General and rooms for the President of the Assembly. If the decoration of the interior is without notable dash or elegance, it is also inoffensive. In general, the color scheme of the working floors is gray—light gray walls, dark gray floors—relieved only by the blue of the doors, the sky blue of the United Nations flag. This, incidentally, is just about the only symbolic indication in the whole building that it has anything to do with the United Nations.

The plan of the working quarters is simple. The elevators, at the center of the building, open on corridors, one on each floor, that run north and south. Since the windows are continuous along the two glass façades, the office space can be arranged and rearranged, when that is necessary, without the interference of external columns; at the moment, the smallest cubicle is two windows wide. Yet even the lucky executives with an ample office at one corner of the structure cannot have the luxury of cross-ventilation, because of the blank north and south outer walls.

The problem of making such a building livable as well as comely was a tremendous one. Let us consider first the matter of orientation. The Secretariat's long axis runs north and south. If the building had been turned so that it was parallel to Forty-second Street, it would not merely have blocked out, for those

approaching the General Assembly Building, the un-
kempt industrial district I have referred to but would
have shielded the Secretariat's occupants from our
intolerable summer sun, which can turn the completely exposed rooms on the east and west fronts
into big bake ovens unless the Venetian blinds are
drawn and the air-conditioning system is working full
blast. Furthermore, the building could, like the Conference Building, have jutted out over Franklin D.
Roosevelt Drive. It would then have been conspicuously visible for great distances north and south, especially when it was lighted up at night.

The engineers concerned estimate that the present
orientation of this narrow building, because of exposure to the sun during the summer, puts a load on
the cooling system that raises the operating cost two
and a half per cent. The human cost is greater; many
of the occupants are compelled to work a good part
of the day under artificial light behind their drawn
Venetian blinds. Thus it is sadly necessary to remove the view the walls of glass were designed to
reveal and to cut off the sunlight they were designed
to admit. Even when the summer heat presents no
problem, there is no possibility of natural ventilation
or a pleasant natural light to work in, despite the vast
amount of window space. Only overhead light, of
course, is really important in an office; a window end-
ing four or even four and a half feet above the floor
would give proper light, and could be opened in
summer to admit a breeze without sweeping unan-
chored papers off the desk. A combination of such a
window and ventilating louvers at floor level would
probably provide a superior system of ventilation.
Why no one has tried this in office buildings (it has
been successfully used in houses) is one of the mys-
teries of an age that boasts of its capacity for experiment.

The result of misorienting the Secretariat and using glass so exuberantly is to create a building that functionally is often windowless on all four sides. On this matter, the architectural historian Professor Henry-Russell Hitchcock seems to me to have said the last word. "The most significant influence of the Secretariat," he recently observed, "will, I imagine, be to end the use of glass walls in skyscrapers—certainly in those with western exposures, unless exterior elements are provided to keep the sun off the glass." In this sense, Sir Joseph Paxton's dream of an all-glass building, which grew out of his experiences as a gardener, has been reduced to absurdity in the Secretariat. The desire for sunlight is sound, for direct sunlight kills bacteria and increases the amount of Vitamin D in the body. Perhaps a window mounted on vertical pivots, so that it could admit sun and air yet be turned so that no capricious wind could disturb a man's desk, is the answer. Lacking such a contrivance, the all-glass wall is an architectural folly, costly in both money and human comfort.

The standard floor of the Secretariat contains offices of various sizes. Because of the shallowness of the building (no point in the interior is very far from the windows, except on the blind ends of the building), this seems an admirable arrangement. Unfortunately, it produces far from ideal working conditions for the secretaries, who occupy the interior offices, where the only daylight is what seeps through the semi-opaque glass partitions that separate the outer rooms from the inner ones. New York's tenements of the eighteen-fifties had exactly the same kind of partition, a pitiful effort to atone for the fact that no daylight or ventilation was provided for the inner bedrooms. To see this symbolic substitute for light and air reappearing in a building that prides itself on its aesthetic modernity is like seeing Typhoid Mary

pose as a health inspector. By opening up the blind
walls, and by moving some of the lavatories that are
unaccountably on the perimeter to the interior, the
designers could have given a certain number of the
secretaries decent exterior working quarters instead
of the stuffy interior chambers they occupy.

In choosing to house the Secretariat in a single
monumental skyscraper, the architects forgot more
than the important matter of maintaining the health
and good temper of the secretaries. They forgot the
necessity of maintaining an *esprit de corps* in such an
organization by means of friendly and professional
contact between people in related departments. At
Lake Success, the United Nations personnel were
housed in horizontal quarters and they went about
the low building under their own steam. Sometimes
they complained about the pedestrian distances, but
they had only to be installed in the skyscraper to realize what an advantage horizontal circulation is in promoting accidental as well as planned meetings and
conversations. Elevator transportation does not favor
this kind of intercourse, and a large cafeteria for mass
feeding does not favor it, either. You would not guess,
from the plan and fitments of this building, that during the last twenty years a much higher regard for
the welfare of the personnel had been making its way
even into hardboiled business organizations, and that
to further efficiency, if not sociability, little breaks in
the day have everywhere become common. A big insurance corporation in New York has actually encouraged the British habit of "elevenses." It is astonishing
to find that the only provision for this kind of refreshment in the Secretariat is the cafeteria. How pleasant
it would have been if the architects had at least provided a solarium on every other floor or so for the
relaxed serving of coffee, tea, and iced drinks. Such

rooms would have given people a destination when they wanted to stretch their legs, which is almost essential for workers as dependent upon human contact and stimulus as the international staff in the Secretariat.

The Secretariat has been presented to the world as a building built about human needs, but that pious profession should bring a blush to the architectural ears of Mr. Pecksniff. Humanly considered, the Secretariat Building is as obsolete as iron dumbbells. The skyscraper form was conceived in Chicago, in the 'eighties, as a means of making the land in the Loop more valuable by increasing the amount of office space that could be piled on it, and it was not long afterward that the tall towers began to serve a secondary function of publicity and advertisement. At no point in the evolution of the skyscraper was the efficient dispatch of business under conditions that maintained health and working capacity the controlling element in the design. Given a magnificent opportunity to break completely away from the stereotypes of narrow-gauge business standards, the architects of the United Nations Secretariat could do no better than perpetuate an outmoded form, save for a few structural innovations that merely decrease its usefulness.

I am aware that I am putting in a minority report on the Secretariat Building. But those who praise it without severe reservations do modern architecture a disservice by judging it mainly on its superficial elegance. A building to house an international personnel devoted to bringing about world cooperation and world peace should be more than a slick mechanical job—and, as I have said, even the mechanical details of this structure are sometimes far from slick. Such a building should, by its zealous attention to human functions and human needs, itself symbolize the great purposes it serves. It should give at least a preliminary glimpse of the new world, the world in which human considerations will be uppermost and will set the mold for all our organizations and institutions. It should be both a visual and an operational symbol, and its beauty should arise out of the due fulfillment of all its functions, graded in the order of their human importance. That is almost the last merit one could impute to the new United Nations Secretariat Building.

1951

SEVEN

United Nations Assembly

The group of buildings that forms the United Nations headquarters reaches its architectural anticlimax in the recently completed General Assembly Building, a fairly low, sort of oblong structure with incurving side walls and a roof that droops in a long, graceful curve between the two ends. This building is the home of what must in time be the most important deliberative body in the world. But there is nothing in its shape, its position, its external treatment, or its relation to the two other United Nations buildings—the Conference Building and the Secretariat—to indicate its importance or that of the organization it serves. The architects who created it would have a hard time defending its exterior even if they had been designing a modern motion-picture palace, which is the only thing it resembles. It is the moving-picture palace of 1950, as the Music Hall was the moving-picture palace of 1930. Instead of a big mural by the conventional Ezra Winter, there are big murals by Fernand Léger, decorative yet equally empty and—is this a heresy?—equally conventional. But as a home for a great institution that seeks to establish peace and cooperation between the nations of the world, it is a painful simulacrum, the kind of thing Hollywood might have faked.

The main axis of the Assembly Building, like the main axes of the two other buildings, runs north and south. Its southern façade gives on the great entrance court that faces on First Avenue, to the west. To the southeast of the building, and joined to it by a three-story passageway, is the low and oblong Conference Building, and farther to the south is the narrow but towering Secretariat. Unlike either of these structures, the Assembly Building has the look of a blank wall from almost every aspect. The only real break in its monotonous expanse of flat stone is the vast window that takes up the whole of the southern end. The side walls were to have been marble, to match the blank north and south ends of the Secretariat, but marble proved much too expensive, and so these walls are only a border of marble surrounding a rather pink English Portland stone, which is not sufficiently close to the marble in either color or texture to blend with it and not sufficiently different to provide an interesting contrast, though time and dirt and the absorptive qualities of this stone may remedy this. Making no attempt to modify these expanses of stone by the application of any detail—sculpture or even inscription —the architects were content to quite literally draw a blank; even the letters "U.N.," which could have been as proudly displayed as the old Roman "S.P.Q.R." or the French "R.F.," are absent. The delegates' entrance, a long, two-level ramp, and a series of exit doorways provide the only accents. (Should one add the minuscule sculptured panels in the doors of the public entrance?) Moreover, there is no approach to the building, for either the delegates or the public, except a sidelong, glancing one. The advantage of a frontal approach, enabling one to see the architectural features, was forfeited by the designers, perhaps for the good reason that they had provided nothing to see. The delegates' entrance, near the southwest corner, is marked only by a panel of marble above the marble-sheathed marquee. The public entrance, at the north end, and as far away as possible from the major subway stops, is equally banal; were it not for an improvised street sign and a modest marquee, one would have no clue to its whereabouts.

These lapses emphasize the consistently undistinguished nature of the structure, whose outer shell gives no clue to either the form of the interior or the activities that take place there. The north end of the building is composed of alternating vertical panels of marble and photographically marbled glass—the pattern of the façade of the Daily News Building, but without its texture or color. This change of emphasis from the horizontal lines of the façades of the other United Nations buildings might have been aesthetically justified if it had been less halfhearted. But the gray featurelessness of these alternating bands of glass and marble is as lacking in character as the bleak side walls; its only effect is to give greater distinction and delight to the treatment of the façade of the Secretariat, as if the architects had looked at the Assembly Building through a reducing glass, for they have eliminated whatever force and dignity even a form as inept as this might have acquired through the use of good detail.

Only one façade of this building, indeed, has any positive architectural quality—the great window, with gigantic square panes, running across the southern end and boxed in by a projecting marble frame. This window dominates, with a certain aesthetic assurance, what one excusably mistakes for the main entrance to both the Conference and the Assembly Buildings. But it wastes its monumentality upon the desert air, for it is not an entrance at all. A small stairway and platform, issuing dramatically from one side of the window, reveal its true purpose; it is a fire exit, and only that. The one part of this monumental structure that visually says "Come In!" is therefore actually saying "Stay Out!" Since this false entrance is of no use for getting in, what purpose does it serve those within the building? This question is not easily answered, for though the view through the great transparent panel is almost unobstructed, the delegates on the ground floor, and the public and the press on the floor above, look out on the sordid industrial wasteland to the south of the United Nations site. Thus a bad site plan, which reveals what it should have taken pains to conceal, is capped by a bad building plan, which compounds this error with interest, in the manner of dwellers on cramped suburban lots whose picture windows face their neighbors' garages and clotheslines.

The three other sides of this building are innocent of any aesthetic decision, and there is little to say of its general form and silhouette except that they tell nothing about the interior. At one stage of the planning, there was a reason for this kind of structural envelope; it was to take care of two separate auditoriums. That accounts for the bulge at each end and for the rise in the roof line. When the decision to build but one auditorium—and ovoid, at that—was reached, the architects had become so committed to this structural envelope that they retained it unaltered, which is all the odder because the auditorium is at the center, the narrowest point of the building. The only outside indication that this auditorium exists is a blisterlike dome of lead-covered copper, too small to be of any positive visual consequence, though it manages to spoil, from certain angles, the one merit the roof could have had, as a sweeping, unbroken curve. Neither functional use nor aesthetic purity can account for this design. Happily, viewed at one point—from the north, along First Avenue—the two visible buildings (the Conference Building is conspicuously absent) suddenly become a vision of delight, when the steep down curve of the Assembly roof, looking somewhat foreshortened from below, intersects the steep marble slab of the Secretariat, to the south. If one could stand permanently at that point, one could forgive all the architectural lapses. Genuine four-dimensional architecture would present a succession of such miracles as one moved around from this point and into the buildings, but the United Nations design, unfortunately, offers only one clean architectural hit. Still, that fine view is worth seeking out, especially on a sunny day with a blue sky, for the pleasures of an abstract composition that is also oddly satisfactory as a symbol of the United Nations itself.

After taking the measure of the exterior of the Assembly Building, one turns to the inside, in the hope of finding some compensatory excellence, but apart from the great lobby and the handsome lounges on the south side, it is almost a museum of modern curiosities. The main public entrance hall has a noble scale, but the details are, so to speak, consistently jumbled, from the marbled glass, whose only intrinsic value is to hide the dirt that accumulates on unwashed windowpanes, to the parabolic arches that support the ramp leading to the first of a series of projecting balconies, a device that reminds one of nothing so much as a Meyerhold stage setting of the 'twenties. As for the balconies, whose billowing forms, finished in white plaster, define the upper levels, they recall the imaginative black-and-white drawings of "plastic" concrete structures that Eric Mendelsohn published some thirty years ago. This treatment is not necessarily bad in itself, but it is surely in aesthetic contradiction to the rest of the structure, and it doesn't belong next to the vertical entrance, with its mulberry-and-blue columns, or the stark, unfinished look of the blue ceiling. This whole entrance hall might well have been left unfinished until the architects could decide upon the perfect treatment for this new type of building. Some

coherent effort should have been made to establish
in the minds of both spectators and delegates that the
nations of the earth have come together here, in all
their variety, their individuality, their richness of historic background, to create a new reciprocity and
unity. As it is, the only recognition of this fact is the
souvenir shop, with its trinkets and minor works of art
from many of the member nations. It is quite plau-
sibly a good political instinct, as well as a taste for
novelty, that draws so many visitors to that modestly
symbolic meeting of cultures. Perhaps what this entrance needs most is a dramatically conceived exhibition space, the sort of thing one finds in a museum of
human geography and ethnography, where the pag-
eant of man could unroll before the eyes. If the hall
were not so narrow and so spectacularly broken up by
the ramp leading from the main floor to the first of
the balconies, and if the spacious side corridors were
not (in glaring contrast to the hall) so low, someone
might still be able to find a way of conveying the meaning and the challenge of this great structure. But the
space has been too badly chopped up. One's attention
is directed to the spatial forms, not to the purposes of
the building. For this reason, the architecture does almost nothing to reinforce the sense of human fellow-
ship and understanding that the visitors bring into the
building.

At the core of the building is the General Assembly
Hall, which is topped by an open-ribbed dome painted
powder blue, with encircling lights pointing down on
the delegates. The south half of this hall, which gives
on the delegates' lounge, is a semicircular wall, slanting
inward toward the dome, ribbed by wood fluting covered with gold leaf, and unbroken except for the
speakers' rostrum and the two continuous side panels
of windows for the radio and television booths. The
other half is open to the press and the spectators, who

are accommodated in a tier of seats rising from the floor and in a balcony above it. Thus this hall is a combination of parliamentary chamber and theater auditorium. A speaker on the rostrum faces both delegates and spectators. Above him is the desk of the presiding officers, and above them is a large bronze shield bearing the United Nations emblem in white, surrounded by large plastic medallions, covered with gold leaf, that will eventually bear the insignia of the component nations. But when one considers that the United Nations has only sixty member countries, the proportions of the hall seem overwhelming. That is not the fault of the architects; rather, as the late Matthew Nowicki, one of the special consultants of the board of design, once observed, it reveals the weakness of the Charter of the United Nations, which, under the formula of "sovereign equality," provides the smallest state with as many seats as the biggest powers. As a result, the Assembly, even while in session, often seems empty, thus defying the conditions Mr. Winston Churchill holds essential to good parliamentary debate. It seems not to have occurred to anyone that this constitution is amendable, and that the composition of the organization could change, in which case even less seating space might be required. Here, as throughout the entire scheme for the United Nations buildings, what is saliently lacking in both the thinking and the architecture is the quality that differentiates modern design from the immobile and ponderous monumentality of the past—the ability to anticipate change and to provide for it. In the Assembly Building, as in the Conference Building, the future is frozen solidly in the form of the present. This lack of flexibility is a serious failure in planning for an institution that may undergo many constitutional changes before it solidifies into a durable mold.

"We were not trying to make a monument," Wallace

Harrison, the architect in charge of the project, has
repeatedly asserted. Whether the architects were trying to or not, they produced a building that has the
weaknesses of a monument—its rigidity of plan, its
sacrifice of function to formal expression. The imposing proportions of the Assembly Hall might have been
justified if the architects had given external expression
to the building through the bold use of a cylindrical
or hemispherical form that would have captured the
eye and perhaps even captivated it. The monumental weakness of this building—or the weakness of its
monumentality—probably stems from the architects'
ambivalent attitude toward the purpose of architecture. In the name of functionalism, they have perpe-
trated formalism, and under the illusion that they were
designing a useful workshop, they have failed to meet
the United Nations' greatest practical need—the kind
of plan that could be adapted to new uses. Despite
this sacrifice, all three United Nations buildings fail
to meet the condition that would justify the subordi-
nation of practical need to aesthetic form—the creation of an endearing symbol of the purposes and
meaning of the United Nations: order out of chaos,
unity out of diversity, peace and harmony out of an-
archic belligerence.

These three buildings do not in any way suggest
in architectural idiom the dawning concept of world
government or make visible the love and cooperation
that are needed for its success. The arid neutralism
of this architecture reflects neither paternal power
nor maternal love; without any warmth of feeling,
without any impressive image of human vitality, these
buildings have only one climax: the thirty-nine-story
skyscraper Secretariat, a type of building that to distant peoples is a stock emblem of the things they fear
and hate—our slick mechanization, our awful power,
our patronizing attitude toward lesser breeds who

have not acquired the American way of life. But this
is the veritable new mother of parliaments, and its
mission is to protect life and nourish life in every part
of the planet, guarding every human being against the
perverse forces that now threaten him. No one should
be able to look at these buildings from afar, or to pen-
etrate their interior, without having his imagination
awakened, his conscience touched, his will to peace
quickened or reinforced, by the design. If the United
Nations matures into an organ of effective world government, capable of affectionately commanding men's
loyalties throughout the planet, it will be in spite of,
not because of, the architecture of its first headquarters.

1953

EIGHT

Workshop Invisible

The General Assembly Building is the last of the group of buildings that constitute the United Nations headquarters. Before judging it I would like to soften my verdict by pointing out the obstacles under which the designers worked. The board of ten architectural consultants (plus half a dozen special consultants) that drew up the general plan was assembled from every part of the world only six years ago. These men were asked to plan a group of diverse buildings that could be put up with the greatest possible speed. If they had been working as a team all their lives, that would still have been a difficult problem, for while they spoke almost the same architectural language, there were serious differences between, say, Le Corbusier, the formalistic French Swiss, and Vilamajo, the genial Uruguayan. Moreover, there were few good modern precedents for the buildings they were called upon to design; modern architecture had had practically no opportunity to deal in monumental buildings, to evolve a truly contemporary site plan, or even to suggest what kind of office building would meet the demands of (in the best sense) bureaucratic organization if anything more than maximum profits was at stake. The only precedents were negative ones—surely *not* the Triangle in Washington or the Pentagon, surely *not* the proposed Palace of the Soviets, surely (though perhaps not so surely) *not* Rockefeller Center. But then what?

Architecture can be produced in a hurry only when the type of building is a well-established one. The lack of good types was a sufficient handicap, but an even tougher problem dogged all who were concerned—the difficulty of formulating a program for these buildings, of deciding, after it had been in operation only a few years, what the needs of this new organization were and how they should be met, considering the limitations of time, space, and money. Today the purely architectural consideration—i.e., the visible structure —is only a small part of the architects' problem. The matter of putting together its mechanical equipment, including such specialties as air-conditioning and radio and television equipment, is extremely complicated, and the prevailing obsession with invention may have inclined everyone involved to sacrifice the permanent function of the United Nations—the pooling of collective wisdom for the sake of world peace—to the novel possibility of dramatically reproducing its procedures by photography and sound and television, though there were voices in the United Nations that warned at the very beginning of the danger of this.

These problems would be difficult even if funds were illimitable and the architects had the leisure of the medieval cathedral builders. But the United Natons demanded, and received, an overnight solution. So these buildings are not the best modern architects are capable of creating; they are merely as good as the group headed by Mr. Wallace Harrison could make them in the brief period allotted. If they are far from timeless in quality, if they are already dated, this is the reason. It is also the reason the architects took an old-fashioned American business building, the product of urban congestion and high ground rents, as their model for the Secretariat Building. In making this choice, they admittedly fell under the spell of Le Corbusier, who, since he apparently has never made a realistic analysis of its workings, regards the tall building as the symbol of advanced architecture. That turned out to be the critical mistake from which many of the other mistakes were derived. The architects sought to fill the eye with a great vertical structure that would dominate the site, and they overlooked their unique opportunity to create a group of buildings that would form a city by itself, visually separate from the chaos around it—a sample of the more cooperative world order the U.N. seeks to bring into existence. Unfortunately, they plunged into their job forgetting that their theretofore useful bag of clichés belonged to a departing era whose social and architectural values were in no way related to these demands.

The United Nations group is approached from the west through a vast entrance court. To the south of this is the United Nations library, a six-story office building that began life as the New York City Housing Authority's headquarters and was later bought by the United Nations and made over. To the east of the court is the Secretariat, a narrow, oblong, thirty-nine-story skyscraper whose axis, like those of the two other new United Nations buildings, runs north and south. The Secretariat, completed in 1950, was the first of these three to open. The second was the Conference Building, a low oblong to the northeast of the Secretariat and facing on the East River, almost the only point from which it is visible. The third is the General Assembly Building, on the north side of the entrance court. This, low, broad, and roughly rectangular, has bowed-in walls and a drooping roof. Though visually the Secretariat stands alone, it is directly connected with the Conference Building, which impinges on it along its eastern flank, and, through a passageway angling around the northeast corner of the entrance court from that structure, with the Assembly Building. I have already discussed the Secretariat *in extenso*. It is my intention now to touch on the Conference Building, a genuinely puzzling structure.

The Conference Building performs three wholly un-related functions: it provides quarters for the councils and committees that run the United Nations, it houses most of the social activities of the delegates, and it serves as a concourse between the Secretariat and the General Assembly Building. This concourse is a three-deck one—a ground-floor level, leading from the Secretariat lobby into the exhibition space and public lobby of the Assembly Building; above this a corridor, for the exclusive use of the delegates, lead-ing to the delegates' lobby in the Assembly Building; and, finally, a level leading into the public lounge and the press lounge in the Assembly Building.

Perhaps the most important factor about the Con-ference Building is that it is almost invisible. One of Le Corbusier's published sketches for the U.N. shows that it was consigned to oblivion at an early stage in the designing. Thus the architects threw away one of the greatest advantages of this site. These seventeen acres afforded an opportunity to design a group of modern freestanding buildings in harmonious aes-thetic relationship. Were the architects afraid to take attention away from their skyscraper, their precious symbol of "modernity"? As if to support this possibil-ity, the part of the three-level passageway visible from the entrance court has, because of its great glass win-dows heavily framed in stone, no relation, in scale or treatment, to either the Conference or the General Assembly Building, the two structures it connects. It looks as if it had been there long before these two were erected and, like the Housing Authority Building, could not be absorbed into the composition.

In plan, the Conference Building is simple. This four-story oblong, sheathed in limestone, is two and a half times as long as it is broad, widening slightly toward its north end, and cantilevered out over Franklin D. Roosevelt Drive. It is surrounded on the second (or Council Chamber) floor by a balcony that juts still farther toward the water and leads, by a flight of stairs, to the still unfinished delegates' garden, which faces a mountain range of confused industrial architecture to the south. On the water side, the building is a seemingly solid wall of glass windows. Behind these windows are the big Council Chambers (Economic and Social, Trusteeship, and Security). Mr. Harrison has said that the architects' only aim was to make these buildings not a monument but "the best damn workshop we could," yet even that limited effort was handicapped by a lack of forthright thinking on matters of comfort, convenience, and function. The immense windows of these chambers do not open, there is no attempt at natural ventilation, there is insufficient provision for easy exit, during a recess, to the balcony for a refreshing turn in the open air. Since the enormous volume of light the windows admit must be screened, even the view of the sky that more functional windows would have given is lost. In the Security Council Chamber, the gift of Norway and designed by a Norwegian architect, the larger part of the window is blocked out by a freestanding screen. Le Corbusier once said that it is now possible to have a window three hundred feet wide, and the United Nations architects seem to have taken him at his word, without asking themselves why, at this particular point in this particular building, an all-glass façade was required. The Conference Rooms, the smaller counterparts of these auditoriums, on the floor below, come nearer to being workmanlike because of their simpler furnishings and their less dramatic atmosphere.

There is an even huger window in the limestone wall of the north side of the building, which houses the delegates' lounge, as well as a bar and a writing room. Here the architects, by using vertical hangings, have broken the glass wall into a more conventional alternation of solids and voids, with a contrast of light and dark sufficient to create a pleasant interior. But they have abandoned the upright wall that surrounds the rest of the structure for a window frame that slants inward from top to bottom, as if this façade were a shop window from which it was necessary to remove any reflection. If one cannot call this art for art's sake, one certainly can call it the cliché for the cliché's sake. The architects have overlooked the aesthetic function of the north façade as a pedestal for the Secretariat, to the south, and as a counterfoil to the almost completely blank curving flank of the Assembly Building, to the west. Black granite might have given the composition the aesthetic vitality the shadowy limestone completely lacks.

As far as its exterior goes, then, the Conference Building is very nearly a nullity. On the inside, I am happy to say, the architects gave a better account of themselves, in part because its furnishings, though modest, have been chosen with excellent taste—rich brown carpets, gray and white walls, graceful chairs and sofas, thanks mainly to Danish and Swedish designers. In the second-floor corridor running south to north on the west side of the building, the architects have used three bowed, freestanding bits of wall to break up what would have been a dreary length of space. The delegates' lounge, too, has been partly broken up by the arrangement of furniture into subdivisions, or wall-less rooms, and a word of approval is due the floor lamps, tripods with inverted lamps and irregular conical shades that give the free-winged effect of a Calder mobile. When the delegates fill this room, it is, in every respect but one, all it should be. The one lack here, and throughout these buildings, is intimate space. Diplomacy probably prospered in the eighteenth century because conversations could be held in recesses of windows behind elaborate hangings. Old London clubs are honeycombed with little private rooms and sequestered lounges. Had this need been recognized in the program, the architects could easily have met it in a variety of ways, perhaps by using a form like the freestanding serpentine wall with which Jefferson adorned the campus of the University of Virginia. Set parallel to the windows, stretches of such a wall would have given the delegates a choice between the dark and the light sides of the serpentine.

In conceiving the building as a "workshop," the architects seem to have thought too concentratedly of television and radio and printing plants and too little about the conditions under which statesmen and their advisers salubriously function. Their bodily needs are well and amply provided for—the Conference Building restaurant, with its view over the river, is charming, and the bar is almost too inviting. But the small room intended for prayerful meditation in the Assembly Building does not meet the need for intimate conversations, for retirement, for withdrawal. Intimate public space is likewise essential—real workrooms, with tables on which papers can be spread out and maps and charts can be consulted. The five committee rooms in the first basement of the Conference Building are too big for this purpose. If the Assembly becomes a real Parliament of the World, the number of standing committees will surely increase, and the need for such space will become even more pressing.

I can find no provision in the building for such expansion short of tunneling farther into the ground.

The interiors of the great Council Chambers are an attempt to do justice to three quite different sets of requirements—those of the members, engaged in discussion; those of the public and press, admitted to the gallery of these chambers; and those of the mechanical organs of publicity, whose radio and television booths take up a whole side of the walls, picking up sounds and images and sending them to other parts of the building and the outside world. The elaborateness of this equipment is fantastic. Because of it and the air-conditioning units, the floors and ceilings of the Conference and the Assembly Buildings harbor a maze of ducts and pipes, complicated to plan and expensive to execute, absorbing the time, money, effort, and imagination that in a simpler culture went into the production of architectural forms. The architect of the Economic and Social Council Chamber—Sven Markelius, of Sweden, one of the original consultants —has even tried to dramatize this unavoidable fact by leaving the rear part of the ceiling exposed. As a consequence, when the delegates' section is lighted up and the public section is in semidarkness, the room looks like nothing so much as a movie set. But when one compares these chambers with such a counterpart as, say, the Directors' Room in the P.S.F.S. Building in Philadelphia, done by Howe & Lescaze in 1932, the timeless simplicity and inner order of that earlier design is evident. The lack of conviction, the element of exaggeration, the recurring touch of uncertainty that pervades the whole design makes even these elegant rooms less workmanlike than they should be. Yet the interior is the only aspect of the Conference Building that deserves to be called architecture. All in all, this structure must be rated a magnificent perversity —a potentially monumental building effectively dis-

guised as a corridor. By missing every opportunity for
good site planning and architectural counterpoint, the
architects managed to wall out of sight one of the three
buildings they were given to design. That would be a
hollow triumph even if this building were—as, alas,
it is not—the "best damn workshop" in the world.

1953

NINE

The Fujiyama of Architecture

New York is finally, to its past discredit and its present honor, harboring its first building by Frank Lloyd Wright, the last and greatest of the trio of master builders that began with Henry Hobson Richardson and Louis Henri Sullivan. This building, a little red brick "prairie" dwelling house, has but momentarily alighted, like a bird of passage, on the edge of Central Park at Eighty-ninth Street and Fifth Avenue; for a few days after these words appear its walls will begin to crumble under the wrecking bar, to make room for the new Solomon R. Guggenheim Museum, which Wright has designed for this block-long site. Next to the house, in an equally evanescent shed, New Yorkers have had an opportunity to look at a comprehensive exhibition —models, photographs, and drawings—of Wright's lifework, spanning the sixty years between the Winslow residence, in Chicago, the first building he did after he left Sullivan's employ and was completely on his own, and his output of the present year. The forerunner of this exhibition, put together under Wright's close supervision by the architect Oskar Stonorov, was unveiled in Philadelphia three years ago. Since then, it has been shown in the great cities of Europe, where it has set in motion a vast wave of homage, accompanied by a platterful of gold medals.

By almost universal acclaim, Frank Lloyd Wright is the most original architect the United States has produced, and—what is even more important—he is one of the most creative architectural geniuses of all time. Today, aged eighty-four, he is the Fujiyama of American architecture, at once a lofty mountain and a national shrine, a volcanic genius that may at any moment erupt with a new plan or a challenging architectural concept or a hitherto unimagined design for a familiar sort of building, such as his recent First Unitarian Meeting House, in Madison, Wisconsin, with its triangular, winglike roof pointing upward like hands folded in prayer. Here, in this exhibit, hung by Wright himself, are his first prairie houses, in which he sought to dramatize the flatness and breadth of the prairie, accentuating the horizontal lines of wall and window, projecting the roof in a wider overhang than anyone had ever used before in this country. Here, too, are his first essays in the skyscraper and the industrial building, which paved the way for the successful buildings of the middle nineteen-thirties.

The visitor's initial impression of this great mass of buildings, which are but a sampling of the six or seven hundred structures he has created, might well be bafflement and confusion. This could be partly accounted for by the inexhaustibility of Wright's genius, his astonishing facility in conceiving new constructional forms and using old materials in fresh ways, as in the tubular fenestration of the Johnson wax company's administration building. He challenges us by risking failure with a new design instead of courting safety—or courting perfection—by refining an old form. This exuberance, this prodigality, is what makes his architecture so difficult for a cowed, security-seeking generation to take; it is much easier for the young architects of our time to accept the consistent tightening and elimination of Mies van der Rohe's work than to enjoy Wright's excess of confident vitality. But the beholders' confusion in appraising Wright's magnificent achievement also lies in the fact that he is neither a qualified historian of his work nor the best exponent of its significance, for the exhibition leaves out a number of clues necessary for understanding his place in the development of modern architecture and for evaluating his productions.

Many visitors to the show seem, to begin with, to have missed the significance of the exhibition house that is part of it. The whole point is that, except for its contemporary fittings, it is in essentials the same house that Wright was building in Illinois almost half a century ago—the Glasner house in Glencoe, the Roberts house in River Forest, the Baker house in Wilmette, a design so sound that even Wright, who scorns to imitate himself as much as he loathes to be imitated, used it in the admirable Willey house, in Minneapolis, which was done in 1934. If today this exhibition house no longer startles, it is a proof of the change that Wright, more than any other architect, has helped bring about—a change in our attitude toward American art, a change from colonial dependence upon European models to faith in our native abilities, from worship of the partly historic to confidence in the living present, from formality and urbane gentility in our style of life to breezy openness and rustic relaxation. Wright has created new architectural forms, symbolized by the use in his interiors of naked brick and stone, unpainted woods, visible beams, cavernous fireplaces. But Wright, magician though he is, was not alone in producing this transformation. Behind it was the whole Romantic movement, which popularized the picnic, the play school, the virtues of country living, and even gave Wright the Froebel toys that, as Professor Grant Manson has brilliantly demonstrated in the *Architectural Review,* so powerfully affected the man's architectural imagination and his system of ornamentation. (His mother not merely

dedicated him to architecture before he was born but
later provided just the right materials for a beginning
architect to feed on.) Even the open plan was not
Wright's singlehanded creation. Thoreau, in "Walden," had dreamed of a house that would be a single
room, in which every utility and every felicity would
be visible as soon as one entered, and the open plan,
in which one room flows into another without spatial
barriers, had been developed, along with the idea of
the horizontal bank of windows, by the Eastern architects who followed the example set by H. H. Richardson in the eighteen-eighties.

The fact that these precedents existed in a diffused,
inchoate form while Wright was still an apprentice
points up the real measure of his contribution. In the
course of fifteen years (1893-1908), he wrought a
decisive change in the form of the dwelling house; he
did away with the cellar and the attic, flattened the
gables of the roof, grouped the windows in long, continuous banks, threw out great overhangs that both
protected the walls and screened out the sun, and, in
every part of the structure, made a consistent if assertive use of all his natural materials. Failing to find
in the market either furniture or fittings that were in
harmony with his new houses, Wright insisted upon
designing these accessories, from chairs and tables to
china and cutlery. A great number of these houses
were boldly Cubist a decade before Cubism, and they
are much better demonstrations of the new aesthetics
that derived from Cubism and the machine than is the
thin two-dimensionalism of Le Corbusier's designs in
the nineteen-twenties.

Even if one does not enjoy all of Wright's dwelling
houses, one must admire the integrity of their logic
and their positive "character." At times the roofs are
gawkily exaggerated and the brick walls dull in texture and color; moreover, in the interest of dramatizing

the horizontal line of the prairie, Wright arbitrarily lowered the ceiling of his prairie houses in a fashion that, under his wide eaves, unduly darkened the rooms and, while it may have reduced heating costs in the winter, also lessened the possibilities of ventilation in the summer. But at other times, he boldly overrode his systematic unification of the part and the whole; in his design for the Coonley house, in Riverside, Illinois, with its isolated rooms and its use of stucco, colored tile, and wood on the exterior, along with great balconies of flowers, he rose from Romantic constructivism to sheer poetry. As a result, even his houses of half a century ago are still fresh enough to be called modern, provided one doesn't limit modern to designate only the clichés of the nineteen-twenties in Europe, a restriction aesthetically absurd and historically false.

One of the reasons Wright's exhibition house has puzzled visitors may be that it has an almost old-fashioned, homey air, as if it had always been part of our landscape. What that means is that many of Wright's most audacious innovations have been generally absorbed during the last half-century; we have taken them in, just as we have taken in glass doors for offices, indirect lighting, steel office furniture—all inventions that stem from Wright or that he helped pioneer. But while his principles were vulgarized in the Craftsman bungalows of the nineteen-hundreds, as they are again in the ranchhouse of today, he himself has never been trapped by his own successes. His gifts as a domestic architect have been demonstrated in the endless variations in plan and elevation that come forth from him in response to a new site, a fresh landscape or climate, a new material, or an untried method of construction. That he is no victim of his own clichés is clearly proved by his houses for the desert, most especially his Taliesin West, in Arizona. Ever since plywood turned his thoughts to the value of circular rooms, Wright's vocabulary of form has been undergoing enrichment; witness the circular plan of his famous San Francisco shop for V. C. Morris in Maiden Lane, to say nothing of a series of residences, among them the David Wright house in Phoenix and the Friedman house in Pleasantville, New York. Wright's constructive inventions and his regional adaptations go hand in hand; his imagination has risen, in Thoreau's words, "to meet the expectation of the land," and as a result no other architect can show such a wide range of regional forms as he has put forth.

All in all, Wright has attempted to wed truth to poetry, nature to the machine; above all, to unite the physical requirements of site and structure to the need for subjective expression, after the manner of his own buoyant, full-blooded social temperament, with its confidence, its optimism, its pride, and its power. The sort of freedom that Sullivan sought to express through his stylized foliate ornamentation Frank Lloyd Wright has infused into the structure as a whole. His strong individual feelings and convictions dominate his constructions; by intention they are neither reticent nor anonymous, for his buildings are corporeal extensions of his personality. This means that to love his buildings, and especially his houses, you must love the man and accept his philosophy of life; you must love him to the point of surrendering to him and welcoming his continued presence in the spirit if not in the flesh. In return for that surrender, he will charm and entrance you, for his fantasy is bold and his sense of the dynamics of space, of the value of contrasting light and shade, of openness and enclosure is superb.

As for the marriage of function with form, Wright never took seriously the doctrine Sullivan preached, that form must *follow* function, nor did he accept the earlier version given out by the sculptor Horatio

Greenough, who equated functional form with the absence of dress or ornament, with economy. In the evolution of Wright's architecture, his tendency has been toward demonstrativeness, toward dramatic exaggeration even when he is at work with the purely mechanical elements. Thus he turns the skyscraper from a cubical cage to an organic form like a plant, with a central core or stem, from which floors, in dramatic contrast to the older system of post-and-beam construction, are cantilevered out—though this more organic design imposes rigidities of its own in fenestration and in the disposition of interior space. Of the two fundamental Freudian types of personality, the hoarding and the spending kind, one tight and compulsive and the other released and generous, Wright belongs firmly to the second. His expansiveness, his exuberance, his inclination to put both form and function highhandedly at the service of his own singular genius are an essential part of his inexhaustible creativeness.

This very quality of personality, so richly interfused in all of Wright's buildings, imposes a great burden upon criticism; one cannot possibly criticize his buildings without making an estimate of his personality, and reckoning with his idiosyncrasies as well as his gifts. As Sir Herbert Read recently said of Wright's latest book, *The Future of Architecture,* "Carried to its logical conclusion, a sense of unity . . . implies that every house Mr. Wright builds is his own house and the people who live in them are not his clients but his guests." Speaking with all reverence for a great master, I must confess that Wright's dwelling houses sometimes put me off by persuading me that he is thinking not of the client's needs but of the architect's own desires and delights. There is a willfulness in taking a plan meant for a city apartment house, built upon a limited space that might justify a hexagonal type of plan and similar rooms, and applying it to a country house. That will-

fulness, for all the charm and aesthetic novelty with
which it is cloaked, is not an uncommon quality of
Wright's work. All great genius shows that tendency in
some degree; it was willful of Michelangelo to place
St. Peter's dome on a structure not designed to support it. But in Wright this quality goes with one that
was perhaps reinforced by the bad example set by
Sullivan. This tendency was one of the unhappy be-
quests of the Romantic movement, for it turned the
artist into a Wagnerian superman, if not a god, whose
intuitions became divine judgments, whose instinctive
preferences become dogmas, whose word finally became law. Wright, fully aware of his own arrogance,
has gaily defended it on the ground that arrogance is
more decent than simulated humility. True, but arrogance is not necessarily better than real humility, the
kind that learns, through self-examination, from its errors, that wrestles with its opponent instead of scorn-
fully dismissing him and so becomes stronger in the
process, that surmounts the limitations of its education and its temperament by seeking to understand
other ways of life, other temperaments, other purposes. Because his own world is so rich, he has little
understanding of how much of other people's worlds
he leaves out.

For all his towering genius, Frank Lloyd Wright is,
by his own philosophy and practice, an Isolato, to
use the word Melville applied in *Moby-Dick* to his
spokesman Ishmael. Each building by Wright stands
in self-imposed isolation—a monument to his own
greatness, towering defiantly above the works of his
contemporaries. Though it dazzles us by its brilliance,
it sometimes fails to invite our love, because it offers
no halfway place between rejection and abject surrender. I hope to expand this matter in the future, when
I propose to deal with Wright's designs and principles
as applied to more collective and public tasks. What is

debatable in Wright's conception of his own role as architect is, as it were, a by-product of that Byronic romanticism, with its inordinate claims for the individual ego and with its contempt for men and institutions that do not conform to it. But this stands in contrast to the best of Wright's work, that which is both most deeply in the American grain and most universal, for he unites the great streams of thought flowing from Europe and Asia into America and creates new buildings and structures in which the modern spirit can feel at home with both nature and the machine, in which every homely activity is transmuted into art.

1953

TEN

A Phoenix Too Infrequent

Because one of Frank Lloyd Wright's characteristic prairie houses has been temporarily set up next to the hut that shelters his show, "Sixty Years of Living Architecture," one might all too easily decide that his major successes are his dwelling houses. That conclusion would overlook some of his most original achievements, and perhaps his greatest potentialities. Except for Bernard Maybeck and Louis Sullivan, there has been no architect in his generation who has had anything like Wright's audacious capacities for handling nondomestic structures. Unfortunately, the growing gentility and middle-aged discretion of American businessmen, even in adventurous Chicago, caused the big opportunities for building warehouses, factories, railroad stations, and public structures between 1893 and 1933 to go mostly to derivative minds, devoted to merchandising Gothic, imperial Rome, or Renaissance hand-me-downs. Rarely has Wright had the sort of chance that H. H. Richardson had to show the full span of his powers. Happily, though, a long succession of projects and buildings, beginning in 1897 with the Luxfer Prism offices in Chicago (never built) and the Larkin Administration Building in Buffalo (razed in 1950), give us more than a hint of Wright's range.

In his larger structures, even more than in his dwelling houses, Wright's university training as an engineer has stood him in good stead. All these big buildings show, in one degree or another, the combination of qualities that makes up his specific genius—his fertility in technical invention and the endless play of his fantasy, largely in relation to new forms provided by the machine. Like sculptors who find heads and torsos and abstract forms in driftwood, Wright quickens in the presence of any building material; he can think of more ways of using glass and concrete, sheet metal and precast blocks, than any of his contemporaries. To respect "the nature of materials"—a phrase often on his lips—and to create original forms in harmony with the mechanical processes that shape them are perhaps his main concerns. These preoccupations override any regard for the varied natures of men whenever they are not in harmony with this effort. Add to this his respect for regular, geometric figures, which, he notes, "have acquired to some extent human significance, as, say, the cube or the square, integrity; the circle or sphere, infinity; the straight line, rectitude." From his willing submission to materials, mechanical processes, and geometric forms, he gets the deepest subjective satisfaction; indeed, to him all this has a moral quality that justifies his indifference to less architectonic human needs and desires. This aspect of his strength partly accounts for a human failing that goes with it: the client he seeks above all to satisfy is himself.

In an age that has released energies and dissolved established forms in every field, Wright's imperturbable readiness to break old molds and improvise new ones has made his less gifted contemporaries look like tired routineers. No one but Wright could have thought of such a fresh departure in store buildings as the V. C. Morris shop, in San Francisco—an all-brick front broken only by a single round-arch opening at one side as entrance, with the sales floors connected by a spiral ramp reaching from the entrance to the glass-bubble skylight. Once he has conceived such a form, he will often reach out for further occasions to use it, and possibly only he would have dared to apply this one to the design of an art museum, like the projected Guggenheim building. When thwarted, Wright seems to lie in wait for a client capable of matching his own audacity and sanctioning the new form. This gives some of his best designs a factitious air when they finally are built. Thus his idea for a skyscraper with floors cantilevered out from a utility core (elevators and the like), the whole to be sheathed in a curtain wall of glass and copper, was originally applied to a projected apartment house for St. Mark's-in-the-Bouwerie back in 1929, but only now has it finally found actual embodiment, in an office building in Oklahoma. About the technical virtuosity of Wright's buildings there can be little argument. His Imperial Hotel, in Tokyo, though wrought of stone and brick without benefit of metal skeleton, has survived more than one earthquake, because it was designed to yield to the ground shudder rather than to resist it. But the infrequency of his commissions for such buildings seems sometimes to have resulted in an infelicitous overcompensation in those that do get built.

The Larkin Building (1905-06), one of Wright's most consummate achievements, will serve as an example. Professor Henry-Russell Hitchcock, so far the main historian of Wright's work, has described it as an expression of "the innate monumentality of an industrial building." As a matter of fact, it was not an industrial building, and its monumentality was at odds with the quiet, direct treatment of its interior. This edifice, far more imposing as a temple than Unity Temple, which Wright built shortly afterward in Oak Park, Illinois, stood by itself in a grimy industrial area of Buffalo. Despite its commercial purpose it had the acoustic properties of a cathedral and the sober austerity of a vast law court, and who could have guessed, on approaching it, that soap coupons were sorted there? What the American office building needed at this point was not greater aesthetic dignity—Sullivan had achieved that—but a layout that would provide more light, air, storage space, and workaday efficiency than the overcrowded skyscraper as yet afforded. Wright's basic plan for the Larkin Building was so sound that it might have shown the way, but his fresh contribution was hidden behind an irrelevant monumentality.

Here is an instance of misplaced creativity, perhaps the first but certainly not the last in Wright's prodigious opus. Denied a sufficient outlet for the imperious demands of his genius, he has too often been impelled to make his own opportunities. So when he finds a client willing to play with him, he has a tendency to project into a building, regardless of economic limitations or functional requirements, all his pent-up creativeness. In the building of the Midway Gardens, in Chicago, in 1913, Wright's elaborate effort to bring together sculpture, painting, and music into an architectural whole (restaurant, beer garden, dance hall) taxed the financial capacity of the owner, and that fact may, almost as much as the coming of prohibition, have hastened its untimely demolition some fifteen years later. In the Larkin Building, Wright expended an enormous amount of talent and energy on a structure whose imposing masses had no meaning for either the dreary site or the business performed in it. In both instances, he may have been overcompensating for the denial of a more normal outlet for monumental building.

Except for a sere interlude between 1920 and 1934, when legal and financial difficulties dogged him and clients lacked the courage to carry out his many proj-

ects, Wright was never a neglected architect. But he
has not ever been sufficiently employed on the sort of
work most suited to his talents—schools, theaters, concert halls, such huge concepts as the San Diego Fair
of 1915, the Chicago Fair of 1933, and the New York
World's Fair of 1939—or on such schemes as the Jefferson Memorial and the National Art Gallery, in
Washington, and the United Nations headquarters, in
which cases the free play of his imagination might have
saved us from a succession of impeccable failures.
Both Wright and his countrymen have lost by this
oversight, his countrymen most of all. The sorry effect
upon his work has been to encourage him, in his lordly
contempt for popular mediocrity, to coddle his idiosyncrasies.

Significantly, when Wright's forms are most strikingly
and unmistakably his own they are sometimes least successful as aesthetic expressions. This is usually due,
not to romantic whimsy, but to a too zealous application of logic. That logic, admittedly, often serves him
well in a simple building, like the Luxfer Prism offices,
in which the client's requirements for uniform ventilation, light, and easily divisible space were translated
into squared bays framed in concrete and filled with
glass. Once such a solution of a specific problem was
found, other designers could take it over, and, as
Wright has acidly commented, they did. But in some
buildings his logic overpowers his aesthetic sense: in
his passion for geometric form, he turns the polygon
into a paragon. Having seized upon a certain geometric form—a hexagon or a triangle—he will, for the
sake of consistency, apply it to every nook and corner
of his design, thus creating a far too insistent series of
harsh, angular forms in the furniture he designs as well
as in the meetings of beams, the shape of his windows,
and the layout of his rooms. One's eye vainly seeks relief from this almost obsessive reiterativeness. This

exaggeration goes hand in hand with an essential trait
of Wright's character: a tendency to replace the engineering principle of least effort with the baroque principle of the greatest show. There are moments when
this even turns into a Hollywood director's love of the
supercolossal. Whether it is quantitative or qualitative,
one must admit that this exaggeration is an American
trait, and Wright, for all his mastery of the domestic
scale, may succumb to its specious enticements even
when they contradict his basic principles of life. Fortunately, the most grandiose of his designs, such as
the gigantic, hive-like recreation center for Pittsburgh's Golden Triangle district, remain in the paper
stage. Wright's best effects are usually simple, as when
he cantilevers the corner of a bedroom, so that as one
opens the casement windows one has an unbroken
view of the garden outside. When he is least showy, he
generally puts on the best show.

In holding an assize on Wright's work as a whole, one
has to ask two questions: what it stands for in itself,
as the unique expression of a powerful and superbly
endowed personality, and what it means in relation to
the time, the place, the community, the civilization.
Wright's methods and concepts, his hopes and ideals,
have not pliantly conformed to the dominant modes
of our time. To what extent are his buildings architectural sports, destined to be engulfed in the chaos of
vulgar building, and to what extent are they prophetic,
as Wright doubtless believes, of a new movement in
modern culture that will transform our environment
and our way of life in a fashion that will increasingly
conform to his own philosophy? Has he struck a note
to which our whole age will in time resonate? (Something like this has already happened in America,
through his transformation of the dwelling house.)
Wright's proposals for the new city and his use of the

machine should provide at least partial answers. In both these fields, we are brought face to face with some of his psychological blind spots, particularly his apparent disinclination, or perhaps his inability, to understand any type of personality but his own, or any way of life but the one he enjoys—an attitude that links him to the antipodal personality of Le Corbusier.

Though Wright got his start in Chicago and his first international recognition in Berlin, he has no love of the city as an entity and reveals little comprehension of either its architectural or its social possibilities. This is not strange; the city is no place for soloists, and Wright, by deepest inclination, is a soloist. That one might need or profit by the presence of other men within an area compact enough for spontaneous encounters, durable enough for the realization of long-range plans, and attractive enough to stimulate social intercourse appears never to have entered his mind. Save for the family, he scarcely recognizes the need for social groups or associations; for him cooperation is a kind of self-betrayal. When he approaches the problem of urban building, he does so contemptuously, and instead of inventing new forms he lowers his sights to the real-estate operators' level. One of Wright's unbuilt projects (Lexington Terrace, Chicago, 1909) would have created what one can only describe as an urban ghetto of back-to-back apartment houses, and his 1940 plan for the Crystal Heights apartments in Washington would have produced just another collective barracks, indistinguishable in the large from other monstrous urban agglomerations. Yet there is an unrealized design of Wright's for a neighborhood community that proves he had real gifts for urban design, had he wanted to develop them. This is the plan for the development of a quarter section (six hundred and forty acres) of Chicago, offered *hors concours* in a competition held in 1913 by the City Club of Chicago. His handling of the open spaces and the public buildings in that plan is so brilliant that it would repay study. But the greater part of the housing area was to be devoted to upper-income dwellings, only four to a block, each house with a corner plot! The possibility that even the well-to-do might benefit by closer association as neighbors, pooling some of their private space in a common, seems never to have occurred to him. By 1932, Wright had turned his back on the city and predicted its disappearance. As a substitute, he offered his plan for Broadacre City, in which the houses of even the least-favored members of the community would stand on an acre of ground. Roughly, this would mean that an area as big as Central Park would hold not more than three thousand people. What is this but the countrywide suburban nightmare that H. G. Wells prophetically pictured in his *Anticipations* half a century ago—at best a somewhat more ordered framework of isolated buildings in which both city and country would vanish, in which one would never get far enough away from one's neighbor to have solitude or near enough to him, without the kind of tedious effort one already needs in Los Angeles, to enjoy the advantages of daily communion or cooperation in common tasks?

This dream of total dispersal, which would carry further the spontaneous disruption of the city that is now going on everywhere, may be only the logical expression of a dominant trait in Wright's architecture: ideally, each building of his must stand alone, free from the support of other buildings, in a completely natural setting. If you put all his structures together in a city, the result would be an aesthetic jungle of dissident, competing buildings; far from lending themselves to contrapuntal treatment, they are all solo performances. To be free, for Wright, means to be free

from one's neighbors. That individuality may reside
in the collective whole, as in a symphony, and that
for the sake of this whole one might willingly surrender some of one's individuality to have it given back,
enriched, at a higher level, is contrary to his philosophy of life and his mode of design. In this inability to
understand either the urban or the urbane, Wright's
attitude discloses the limitations of Romanticism, with
its rebellion against everything that demands conform-
ity to a general social pattern. To the problem of
bringing individuality, personality, spontaneity, freedom back into the huge mechanical urban collectives
that now operate so compulsively, he has no answer
except "Clear out!" Thus he has never faced the para-
mount problem of modern architecture—to translate
its great individual accomplishments into an appropri-
ate common form in which, by pooling economic and
social resources and cooperatively integrating designs,
advantages that are now open only to a wealthy few
will accrue to a great many. If the contemporary architect has not as yet found an adequate answer to
this problem, Wright characteristically has not even
asked the question.

As for the machine, Wright's approach to it has been
ambivalent, not to say paradoxical. Though he was
possibly the first modern architect to freely accept the
machine, in ornament as well as construction, he has
little use for its indigenous forms—the impersonal, the
typical, the anonymous. Le Corbusier gave a fresh im-
petus to the modern movement by showing how much
good form had already been produced by the machine, in ordinary drinking glasses, pipes, bentwood
chairs, and office equipment; Wright, on the other
hand, saw machine production as a way of producing
new forms that would bear his unmistakable mark.
So his furniture, specially designed for his houses, has
fitted only his own unique architectural plans, both at

the expense of the human carcass—he admits he always barks his shins on his chairs—and at the expense of the influence that a more neutral design might have exercised on furniture in general. Wright's dislike for the typical and the generic is probably also the explanation of his hostility toward those who have won favor by using less personal forms. But surely one of the reasons for the appearance of the cardboard-box style of architecture, which he so violently detests, was the desire, in a world full of assertive advertising and *art-nouveau* idiosyncrasy, for the almost monastic simplicity and anonymity of Cubism and purism. Wright, it is true, had no need to participate in that revolt, for he had kept away from *art-nouveau* curlicues as rigorously as he had from moth-eaten historic furbelows. Indeed, in many of his earlier houses, which were disciplined by his love for Japanese forms, he had achieved a similar cleansing, a similar clarity, in his own right, long before Ozenfant, Le Corbusier, Mondrian, and Gropius had discovered abstract art's antiseptic charm.

How, then, is one to account for Wright's hostility to the "box," a hostility that seems to deny that it is one of the eternal, basic geometric forms? Is he not chiefly offended by the fact that its neutral background might allow for intimate choices and delicately personal needs —for a picture or a statue, for a set of fine chairs from the eighteenth century—that would be put out of countenance in a room with a more positive aesthetic character? Probably the answer for Wright is that the room itself, as molded by the architect, is the personality to be considered, not the user (he is as insistent as Le Corbusier that the client should replan his life to fit the new structure), and if this means discarding cherished possessions, giving up pictures on the wall, accepting the natural colors and textures of the materials used, it is for him a better solution than the toler-

ation of architectural anonymity or neutrality. Wright
has identified his personal philosophy with "democ-
racy" or "the Usonian [Wright for "American"] way of
life," offering the implication that America must guard
itself against all manner of foreign importations, especially architects. More than once, in expressing his
opposition to the "box," he has spoken as if the work
of his foreign-born architectural rivals were not merely
"inorganic" but un-American. While Wright is sound in
asserting the American architect's freedom from colonial servility, it is another thing for him to denounce
architects of European origin, such men of integrity
and humane understanding as Gropius, in language
(and thought) that should be reserved for morbidly
isolationist journals. The America First streak in
Wright is a coarse, dark vein in the fine granite of his
mind, and it has kept him from learning as much as he
might have from those who by taste and temperament
and training most differed from him.

The proportions of this loss become clearer when
one discovers that some of Wright's most distinguished
buildings are those in which, consciously or uncon-
sciously, he started from the same general premises
as his foreign rivals, and in which he used their vocabulary, though admittedly with more eloquence.
Take his Johnson's wax company tower, in Racine,
Wisconsin, with its smooth bands of brick, rounded
even at the corners, alternating with smooth windows
of glass, without the slightest expression of the third
dimension except for the shadowy forms within—
what is this but an outplaying of Mies van der Rohe
at his own game? Or take the building that is by critical consensus one of Wright's masterpieces, the Kauf-
mann house, at Bear Run, Pennsylvania, dramatically
poised over a waterfall—what is this but a transla-
tion into three-dimensional concrete forms, with
thrust and dynamic counter-thrust, of the structural

elements that Le Corbusier so often reduces to a thin, two-dimensional envelope? When Wright is most susceptible to such discipline, whether offered by a client or a rival, his form is often most effective. His true genius, like the true genius of America, is expressed in neither his rejections of the foreign nor his self-assertions of the merely native but in his generous comprehensions and inclusions, his outgoingness and his receptivity, his unrivaled capacity for synthesis. The machine he celebrates was a gift of Britain, whose builders used glass and iron a generation before the designers of the first Chicago skyscraper; his cult of nature was a gift of the French philosophers and poets of the Romantic movement; his most stimulating education came via Froebel, from Germanic sources, like Bach and Beethoven, his lifelong loves. As for the influence of Japan, his readiness to accept it was an expression of the union of the East and West that began long ago in the minds of Thoreau and Emerson—a contribution to that One World from which Wright, in his more ingrown moments, shrinks. This receptivity, this universality have given the United States a special position as a sort of Grand Central information booth, a public meeting place where many races and many minds may freely exchange their best thoughts, perfecting their understanding of each other and thus increasing their tolerance, sometimes exalting their love.

This touch of universalism (for it is still only a touch) is what distinguishes American nationality from the more orthodox kinds. When Wright's words about American architecture—imputing to his own "Usonian" architecture a monopoly of the democratic virtues and of organic design—seem to betray him, one must remember the meaning of his whole opus. The speech he has put into steel and glass and stone is a living refutation of his isolationist philoso-

phy. What is great in his genius is perhaps what he
values least—not his Americanism, not his Romanticism, not his originality—but his universality. The
fact remains that in a period of specialist constrictions
and nationalist conformities his lifework has expressed
the full gamut of the *human* scale, from mathematics
to poetry, from pure form to pure feeling, from the
regional to the planetary, from the personal to the
cosmic. In an age intimidated by its successes and
depressed by a series of disasters, he awakens, by his
still confident example, a sense of the fullest *human*
possibilities. What Wright has achieved as an individual
in isolated buildings, conceived in "the nature of materials," our whole community, if it takes fire from
his creativity, may eventually achieve in common designs growing more fully out of "the nature of man."

1953

BLIGHT AND BEAUTY

ELEVEN

Skin Treatment and New Wrinkles

The last six years have brought forth a boom development in midtown office buildings in Manhattan. With the exception of Lever House, these buildings have increased the congestion of New York's midtown business areas without adding much to the glory of its architecture. As a whole, they are an ironic commentary upon the intelligence of our financiers (vestigial) and the civic foresight of our city planning commission (absent). The one creditable fact about these new human filing cases is that the loftiest of them is only twenty-eight stories high. This limitation of height, which is an at least partial rejection of the old cathedral-tower type of building, may mean that investors are belatedly beginning to assimilate the elementary economic lessons of tall buildings. The relation between height and rate of financial return was set forth thirty-odd years ago by the Chicago architect George Nimmons, who designed the early Sears, Roebuck department stores. Nimmons, basing his study on typical Chicago office buildings, demonstrated that the maximum possible rate of yield on the cost of a thirty-story building was less than that on a ten-story structure. A twenty-story building gave the highest yield (7.05 per cent), but the yield on a fifteen-story building was only a quarter of one per cent less. Nimmons entitled his paper "The Passing of the Skyscraper." He called the turn a little too

soon—probably because there is a difference between putting up a building for immediate profit and
putting up one for a permanent income. A building
that can mean a big reward for the builder who over-
crowds a site can mean a low income for the man
who buys it from him as an investment.

Only the new directors of the New York Central,
who lately have been dreaming of hundred-story office buildings, seem adamant today in the face of
these facts, and thus the supercolossal no longer
threatens to completely dominate the midtown New
York horizon. Unfortunately, though, supercongestion
tion remains, for the height and density of population
of the new business buildings are just about double
what they are in the ones erected along Madison Avenue and Park Avenue in the busy years between
1914, when construction activity near the then new
Grand Central Terminal began to quicken, and 1924,
even though the city ordinances of 1916, which controlled the size of these structures, were made more
stringent in 1943. As a result of these recent encroachments, the splendid scale of Park Avenue set
by its early hotels and apartment houses is disappearing as one substantial building after another gives
way to a new pile of offices. There is no longer anywhere in the heart of the city as much as half a mile
of buildings—Park Avenue once boasted two or three
consecutive miles—that combine coherence and civic
dignity. And I don't know which must be counted the
more lamentable—the fact that this change has taken
place, or the fact that so few people seem to realize
what has been lost in this rebuilding process.

The mischief the city is now facing in midtown
Manhattan is twofold—first, the density of occupation is increasing, and thus increasing the density
of traffic; and second, this area, which, because of its
proximity to the big railway and bus terminals, as well

as to the amusement and shopping district, is the natural residential quarter for visitors, is being disrupted by the growing invasion of office buildings. This shift in purpose has been accelerating in the past few years. But the reasons for the congestion of the area go back forty years, to the ordinances of 1916. Those ordinances, a pioneer effort at city planning by force of law, established the principle of dividing the city up into zones, or areas, dedicated solely or principally to business, industry, or residence. Likewise, they recognized the necessity of controlling the height and the bulk of buildings.

At that time, those sufficiently ancient may remember, Madison Avenue and Lexington Avenue, above Forty-second Street, were still flanked mostly by four-story brownstones (many of them, it is true, in process of being converted to commercial uses), and Park Avenue was a waste of railroad tracks. Except for the Times Building, there were almost no tall office buildings north of the Flatiron Building, at Twenty-third Street; indeed, the bulk of commercial building was still below Park Row. It was the reckless piling up of high structures near the southern tip of Manhattan, cutting off each other's light and air, that prompted these ordinances that sought to limit the height of buildings and the area covered by the upper floors. Because the population then relied far more on heavy-density public conveyances than on the light-density motorcar for transportation, the need for controlling the density of occupancy in buildings, which is the basic method of controlling traffic congestion, was not understood. That should not, however, surprise us. The failure to understand this relationship is today practically an indispensable qualification for setting oneself up as an authority on traffic problems.

The officials who devised the ordinances were men

not without civic vision, but they allowed that vision
to be clouded by a mixture of commercial considerations and romantic aesthetics. When the early planners drew up the ordinances, they did not take into
account the rather low land values that prevailed in
the all-but-virgin midtown area (low land values
make low buildings economically feasible), nor did
they grasp the opportunity to establish more desirable
conditions than prevailed in the portion of the city
that had already been overbuilt. Instead, they perpetuated the very problem they were trying to solve.

The principal provision of these ordinances as they
applied to the Grand Central area, officially ruled a
business and not a residential zone, was that the façade of any structure could not rise vertically from
the building line to a height greater than twice the
width (from building line to building line) of the
avenue on which it faced. If a structure was taller than
that, the stories above this point had to be set back
so that no portion of them projected beyond the extension of an imaginary line drawn from the center
of the avenue to the point at which the first setback
began. (The 1943 revisions reduced the façade height
to one and a half times the width of the avenue.)
On the side streets, less stringent regulations prevailed. The size of the setbacks naturally increased
as the building increased in height, bringing about the
terraced, pyramidal look now familiar to Manhattan
cliff dwellers. These setbacks did enable a modicum
of light and air to filter into the newly created city
canyons, but at the same time they enabled the builders of office buildings to vastly increase the block-
by-block population of this district. The fact that
they were only partly effective from a functional point
of view was supposedly mitigated by the fact that they
encouraged the architect to conceive a novel type of
skyscraper, whose setback upper stories made it re-

semble a Babylonian ziggurat. But why Babylon? And why a ziggurat? Perhaps it was the same line of reasoning that had caused the old Tombs and the reservoir that once occupied the site of the Public Library, at Forty-second Street, to be Egyptian. Perhaps someone had recently happened to read a new book about Babylon.

In short, the laudable principle of height and bulk limitation was established, but though they established it, the lawmakers surrendered important points to the innate cupidity of speculators and investors. The compromises they made nullified the main purposes of this excellent law, and they were never sufficiently corrected, even after the principle of regulation had run the gantlet of the courts and had been ruled a proper concern of a municipality, inasmuch as it had a bearing on public health and welfare. For some reason, the earlier office and apartment buildings on Madison Avenue and Park Avenue did not go as high as the law allowed—a good thing, too, since for some even stranger reason the law was more lenient about the height and size of buildings in this midtown area than it was about buildings in most other parts of the city. Though these office buildings, in contrast to Lever House, covered their whole plot and created what was, before air-conditioning, an intolerable amount of airless inner space, they were at first built only to a height of twelve stories. During the 'twenties, however, the roof was, so to speak, lifted by supposedly sober businessmen who seemed determined (since they were encouraged anyway by the leniency of the law) to build higher than their neighbors, even at the expense of profits. The architectural effects of that competition were lamentable. With the exception of such cleanly setback towers as the Empire State Building, the ziggurats, in their effort to enclose in their setbacks every cubic foot of space the law allowed, were often askew on their off-avenue sides, and if by accident they achieved any beauty, it was soon hidden behind an even taller neighbor.

Oddly, there has been no fundamental advance in skyscraper design since the eighteen-nineties. The present most fashionable model, the tall, thin slab, exemplified by the United Nations Secretariat, by Lever House, and by a new structure at 430 Park Avenue, goes back to Burnham & Root's design for the Monadnock Building, in Chicago (1889), which was sixteen stories high and had light and air on all four sides. The slab form is an ideal one for providing light and air, but it is as outmoded as a Roman colonnade in a day when air-conditioning and effective artificial indoor lighting are commonplaces. The one positive contribution the slab has made in our time is that it has released the imaginations of architects and investors from the spell of the zoning-law ziggurat and the earlier, pseudo-cathedral tower of the Woolworth Building period. So my report on the latest batch of office buildings must be confined to superficial matters, for the important changes are only skin-deep, and the greatest success has been in the trying out of novel constructional wrinkles.

Take the new office building at 99 Park Avenue, designed by Emery Roth & Sons. This may be looked upon as the last word in design as envisaged by the framers of the 1916 ordinances. It occupies the east block front on Park Avenue between Thirty-ninth and Fortieth, and runs back five bays toward Lexington Avenue. This is a plot big enough to have encouraged its owners to erect a slab, but its very size seems instead to have persuaded them to "fill the zoning envelope," as the process of using every cubic foot the law allows is called. (The city ordinances, which are nothing if not complex, permit the erection of a slab building to any practicable height if the ma-

jor portion of the site is built up in accordance with
the zoning rules.) Happily, the fact that the site of
No. 99 is a block front makes it possible, under the
regulations, to step the upper stories back in symmet-
rical fashion, so it is an almost ideal example of the
commercial ziggurat. But the main new feature is its
outer shell, or its epithelial layer, which is of aluminum sheathing. In this building, the process of aluminum sheathing, first tried in the window frames of the
Empire State Building and then developed into a complete system of paneling by Harrison & Abramovitz
when they designed the Alcoa Building, in Pittsburgh,
has been carried a step further. No. 99's prefabricated panels, two stories high, consist of two aluminum window frames and two spandrels, which cover
the fireproof brick fill between the floors. The spandrel panels are stamped with a geometric figure whose
indentations give additional rigidity to the thin aluminum sheet.

From a distance, this looks like the usual skyscraper
of our period, with the windows, four to a bay, forming almost uninterrupted strips around the façade.
Fortunately for the window cleaners, however, these
windows are set on vertical pivots. For this feature
alone, the design deserves an award from the National
Safety Council, and if the building were not air-
conditioned, I would also give an honorable mention
to these windows, for, if opened, they would permit
the maximum amount of sunlight to enter the interior
without screening out, with their glass panes, most of
the salubrious sunlight. Apart from this, the principal
merit of the sheathing seems to be facility in both
construction and destruction. Three crews set the
eighteen hundred window panels in six and a half
working days. (One must not forget, however, that
the brick backing of this surface was still constructed
in the usual manner and required the usual time to

construct.) This metal covering, furthermore, doubtless gives efficient insulation and weather protection,
provided the joints hold under severe changes in tem-
perature, which always put a strain on metal construction.

The aesthetic advantages of this new sheathing
are not, however, unmixed. Aluminum, for all its
luster, is a sober material. The faceted panels give
effective contrasts of light and shade that had almost
been overlooked in the recent spate of slick, unorna-
mented surfaces, and there is likewise a visible difference of tone in the façade, shading from light gray at
the top to dark gray at the bottom. This effect is
purely an accident, not an intended result, for the
light is naturally dimmer at the bottom of our city
canyons than it is at the top. A whole avenue of aluminum walls would be dismal, and as grime overlaid
the surface, it might likewise become dingy, too,
though not beyond the remedy of steaming and scrub-
bing. For Pittsburgh, a grim, masculine, Scots-Irish
sort of city, whose darkness magnifies the flare of the
furnaces, this dour quality is not inappropriate, but in
New York, which has hitherto lifted a bright, almost
feminine face to the sky, this material can be wel-
comed only as an occasional note of contrast, comparable to that provided by the modern phenomenon
of the dark-hued all-glass façade. One understands
now why Frank Lloyd Wright, in planning a skyscraper, decided to sheathe it in copper. The other
defect of this aluminum skin is that since it is only an
eighth of an inch thick, it tends to develop a slight
surface wave. All in all, the financial advantages of
this mode of construction seem to outweigh the aesthetic ones.

                                              1954

TWELVE

Fresh Start

The announcement that a big local manufacturer has
built a plant in the Erie Basin area of Brooklyn is not
in itself enough to suggest a pilgrimage to that particular benighted district. There are plenty of low, modern factory buildings in Queens and Brooklyn, and
even the news that this one has seventy-five thousand
square feet of floor space on each level does not seem
important. What makes the new plant significant is
that at last one maker of men's clothing, Eagle Clothes,
Inc., has defied the holy principle of adding to New
York's intolerable congestion and has deserted overbuilt midtown Manhattan. What is more remarkable
is that this breakaway was encouraged by the Amalgamated Clothing Workers of America, which for years
has apparently feared any dispersion of the garment
industry.

Let us deal first, however, with the purely architectural aspect of this project, since it isn't especially
vital. The building (at 225 Sixth Street, between
Third and Fourth Avenues) is a solid, two-story concrete structure, faced with orange brick and opened
up on all four sides by long strip windows. Aesthetically, it would rate as undistinguished but decent except for a tall, raucous sign announcing that this is
the home of Eagle Clothes. Because the building is
as broad (two hundred and seventy-three feet) as it
is long (two hundred and seventy-five feet), the windows contribute little to the interior lighting. This is

mostly taken care of by rows of fluorescent fixtures
that run the length of the factory. The office section
and the first-aid clinic, appropriately, are on the
sunny south side of the building, but the big cafeteria
in the basement (which, by the way, serves break-
fast as well as lunch) is almost an afterthought, for
the windows are little better than transoms set close
to the ceiling. Because of the ample width of the aisles
and the wide separation of the columns, the two busy
manufacturing floors have an air of clean, uncluttered
serenity. Though not shallow enough for natural lighting, these are sound working quarters; technologi-
cally, they are just about all that a manufacturing plant
should be. On the whole, both men and machines are
treated well.

The advantages of this factory are plain. It has
been put up where the nondescript industrial area of
the Erie Basin, leaping across the Gowanus Canal, begins to give way to the modest residential streets of
South Brooklyn. Thus the makers of Eagle Clothes
have traded the doubtful luxury of height and prestige in one of the cramped loft buildings of Manhattan's garment district for the blessings of ample horizontal space, for on land as cheap as this a factory
can be laid out in its most efficient form—low and ex-
pansive. The trucks that bring in the bolts of cloth
and take away the boxes of finished suits do not waste
time finagling for curbside space. Two subway lines
are close at hand for employees who do not live near
their job, and on pleasant days, if the lunchroom is
hot or crowded, they can eat on the roof in the open
air, not at the bottom of a canyon but in full sunlight.
When the wind is in the right quarter, that air is fra-
grant with the smell of coffee from a nearby roasting
plant. Even Manhattan benefits by this removal; because nearly a thousand workers are taken off its
crowded streets, there is that much more elbow room.

Not a great deal of elbow room yet, but if a hundred other factories were to follow suit . . .

This is a case of individual initiative in the right direction, initiative that should be handsomely recognized and taken up on a much wider scale. No part of the city cries more loudly for reorganization than the Erie Basin and Gowanus Canal district of Brooklyn. It is a region of grimy factories and warehouses and gas tanks, of badly paved, narrow, abortive streets, of empty lots and industrial rubble among which gnarled trees and abandoned privet hedges manage to survive. It looks like a segment of a bombed city. In the midst of this emptiness, the Brooklyn Improvement Company, whatever that may be, occupies a classic stucco mansion, standing at the corner of Third Street and Third Avenue in ironic solitude—or should one say hopeful anticipation? Viewed at twilight, this wasteland becomes a romantic townscape, given a striking accent by the great arcs of viaduct that sweep across the sky, carrying the Independent subway trains and the Belt Parkway, which bypasses this area to the west. Here is one place to stake out an industrial redevelopment area, with an eye to taking such basic New York trades as garment making and printing out of the midtown area and settling them in low factories in a new kind of industrial district, as spacious and handsome as the planned industrial areas in England, such as the North Eastern Trading Estate, in the Team Valley. The complete confusion of the area presents serious obstacles to redevelopment, but the territory to the east of the Gowanus Canal, up to Fourth Avenue and the Eagle Clothes factory, is ripe for conversion into a light-industry center. Part of this section is dump heaps and vacant lots, and part is occupied by dingy two-story brick dwellings and even dingier four-story tenements, embellished mainly with TV aerials, that belong to the dark ages of housing. This is just the

sort of run-down neighborhood the New York City
Housing Authority might pounce on for one of its skyscraper projects, but the good traffic routes to downtown Brooklyn and to New York (via the Belt Parkway and the Battery Tunnel) make it the logical
place for light industry, and any new housing should
be kept farther east, beyond Fifth Avenue.

Such an industrial area cannot simply be zoned; it
must be planned in detail—with wider arterial thoroughfares and fewer cross streets, off-street loading facilities to relieve street traffic, and the elimination of
all unfit buildings. The new factories should be low
and grouped around small parks lined with shade
trees, in which the workers could take their noontime
and after-hours recreation. What one manufacturer
perhaps cannot afford to do by himself, a group of
manufacturers could easily do by participating in a
common plan—for which the eminent-domain powers
of the state could be invoked, as they are used to im-
plement a housing redevelopment scheme—to buy up
and replan as a unified whole a sufficient amount of
land. And what the Amalgamated union has done during the last generation to improve the worker's life
within the factory should be paralleled by an effort
to humanize a whole industrial quarter through imaginative planning. In the garment industries, as it
happens, the worker's recreational life provided by
the union is to a large extent tied to his place of work
—lectures, study groups, dramatics, and dancing—and
there is no reason for that environment to be depressing. A well-planned industrial quarter might do
more for the worker than any amount of psychological
spoon-feeding.

Every scheme for relieving traffic congestion in the
central areas of the city will fail until this kind of decentralized redevelopment takes place. More than a
generation ago, that farsighted businessman Irving

Bush made the first try when he put up the vast Bush Terminal along the South Brooklyn waterfront, where rail and water transportation and efficient manufacturing facilities were, so to speak, brought together under one roof. Neither his business contemporaries nor the municipality, unfortunately, had the imagination to follow through; business as usual still provided pickings that were much too good. But now that New York is strangling itself, perhaps a few people besides the makers of Eagle Clothes will try to unknot the noose before the blood completely stops circulating.

1952

THIRTEEN

Prefabricated Blight

There is now, within New York's city limits, a good handful of housing projects in various stages of completion. Since the housing shortage is what it is, any truly critical appraisal of these undertakings will have to come, I am afraid, from a nontenant. Those who are lucky enough to be accepted by the landlords will probably view their new homes through a rosy haze, as a Displaced Person might view Ellis Island. Yet almost all these projects are solemn reminders of how different the postwar world is from what most people hoped it would be.

Perhaps the greatest and grimmest of these housing developments is Stuyvesant Town, the massive palisade of apartments being erected by the Metropolitan Life Insurance Company between Fourteenth and Twentieth Streets, in the area bounded on the east by Avenue C and a small stretch of the Franklin D. Roosevelt Drive, and on the west by First Avenue. By the time this group of buildings is finished, twenty-four thousand people will be living on its sixty-one acres. (The original street and block pattern of New York was planned for a density of from seventy to ninety people an acre; this new development will have a density of three hundred and ninety-three.) That makes this community larger than sixteen thousand other towns in the United States and smaller than only four hundred. Instead of lowering the density in the area, the proprietors of Stuyvesant Town, abetted by the City of New York, have established a pattern of greater congestion. If New York were completely rebuilt in this fashion, even Mr. Robert Moses, who has had a lot to do with setting this pattern in housing, would perhaps cry "Uncle!"

When I first inspected Stuyvesant Town, a year ago, the development seemed to me an unrelieved nightmare. Though the buildings are not a continuous unit, they present to the beholder an unbroken façade of brick, thirteen stories high, absolutely uniform in every detail, mechanically conceived and mechanically executed, with the word "control" implicit in every aspect of the design. This, I said to myself, is the architecture of the Police State, embodying all the vices of regimentation one associates with state control at its unimaginative worst. But the entrepreneurs of this enterprise are not commissars; they are the president and directors of a great life-insurance company, and they have performed this feat of regimentation in the name of free enterprise and individual initiative. To encourage the exercise of these virtues and to coax private capital into this large-scale housing effort, the City of New York aided in the acquisition of the land and allowed the new owners to wipe out the existing network of streets, which occupied over twenty per cent of the area, without exacting any payment for the additional land thus acquired, although two small strips, along First Avenue and Twentieth Street, were ceded to the municipality so that it could widen these thoroughfares.

In order to make the Metropolitan's control over this feudal domain absolute and inviolable, a public school and two parochial schools in the territory were also demolished. Now no citizen of New York may set foot within Stuyvesant Town except by permission of the owners, and a private police force is on duty to exercise, if the proprietors require it, this control. And that is not all. Principally, I believe, at the urging of Mr. Moses, as City Construction Coordinator, the city has consented to waive all taxes on the improvements for twenty-five years—amounting to an estimated fifty-three million dollars—to insure housing at seventeen dollars a room per month.

There is no provision in the agreement between municipality and owners to provide housing for the people driven from their homes by the clearing of the land, nor is there a provision that only low-income groups shall have the benefit of this very generous subsidy. Naturally, the Metropolitan Life, like any other prudent landlord, is selecting its tenants on the basis of "desirability," from a group of people who, though mostly veterans, are a fair distance from the ragged economic edge; the construction of six large underground garages for them is evidence enough of that. Only in the Looking-Glass World of Lewis Carroll does any of this make sense.

A year has passed since I saw the first units of Stuyvesant Town, but after a second inspection I find that my nightmarish impression of the project has not dissolved. As things go nowadays, one has only a choice of nightmares. Shall it be the old, careless urban nightmare of post-Civil War New York, planned so that from a third to a half of the interior space of each house and apartment is dark, airless, dismal, so that there is no view worth speaking of from the windows, except by accident, and so that there is a maximum amount of noise from the streets? Or shall it be the new nightmare, of a great super-block, quiet, orderly, self-contained, but designed as if the fabulous innkeeper Procrustes had turned architect—a nightmare not of caprice and self-centered individualism but of impersonal regimentation, apparently for people who have no identity but the serial numbers of their Social Security cards? Since I doubt that the directors of the Metropolitan Life Insurance Company really wished to produce this effect, it might be interesting to see how it all came about.

Let's look first at the over-all plan of Stuyvesant Town. To understand that plan, one must visualize the basic unit—an elevator apartment house in the shape of a cross, with four wings of equal dimensions. There are four such units on First Avenue, and three others on Avenue C. But mostly two, three, or four units are joined together into rows, in which there are occasional variations, to round a corner of the site. These rows form, roughly, a series of open quadrangles around the perimeter of the super-block. (Local traffic moves within this area along curving driveways, and curving walks provide pedestrian access. The space wasted by streets in the typical old city plan is here fortunately converted into playgrounds and open space.) The rest of the units, eight in number, form the spokes of an imaginary oval wheel, splaying out from a large green, called Stuyvesant Oval, at the center of the plot. This oval is planted with big oaks and surrounded by a triple row of plane trees, all of mature growth. Except for a few skyscraper apartments, which temporarily have uninterrupted views from their upper stories, nowhere in the more expensive residential districts in Manhattan will you discover any comparable access to sunlight, air, and vista. Nowhere else, either, have the differences in level at various parts of the site served better for handsome plantings of rhododendron, with dark ivy ground cover to contrast with patches of greensward.

Superficially, all this seems like a positive gain, but it will not bear close examination, for neither functionally nor visually have the planners made the most of

their site. Far less land in the area is now occupied by
streets, but, on the other hand, all the major open
spaces except Stuyvesant Oval are given over to completely asphalted playgrounds, surrounded by spiky
iron fences and paved walks, that offer none of the
natural resources or the implements for any kind of
play except the most mechanical and organized. The
needs of different age groups are provided for only by
differences in the type of playground equipment. Since
most of the families in Stuyvesant Town belong to
young veterans and the small fry are almost all under
five, the woodenness of the scheme is not yet fully
apparent, but once time produces the normal variety
of children, from toddlers to adolescents, this failure
will become glaring, and the alternative to conflict
and frustration will be more regimentation. However
they are labeled, the open spaces in this development,
it is evident, were not designed specifically at the beginning, as they should have been, for special uses.
There should be hedges and windscreens for mothers
and infants, digging pits and water pools for the toddlers, and trees and bushes for hide-and-seek and
provision for games of make-believe for still older
children, and each space should show, by its size and
shape, the function it aims to serve. Actually, these
playground spaces are merely leftovers, bleak asphalt
wastes, marks of that absence of human interest, of that
almost positive distaste for beauty, which characterize
this project. Not merely the human scale but human-
heartedness is lacking.

If you did not know what the ultimate population of
Stuyvesant Town would be, you might think that the
provision of open spaces is exceedingly generous, but
when the density of population is taken into account,
it is plain that what has been allotted does not meet
even the most meager requirements of good planning.
City planners everywhere are agreed that one acre of

open green space to every hundred people is desirable and that one acre for every three hundred people is the barest minimum standard for health and decency. On this minimum basis, the population of Stuyvesant Town would require, apart from the normal space between buildings, some eighty acres of open land, which is nineteen more than are taken up by the entire project. So, instead of easing the general deficiency of parks and playgrounds in Manhattan—in return for the city's munificent subsidy—Stuyvesant Town actually adds to this chronic shortage.

Visually, the case against this development is just as strong. By the exercise of extraordinary ingenuity, the landlord's Board of Design—let its members remain anonymous—contrived to accentuate the stereotyped character of these buildings by so placing them that one cannot anywhere find a vista that is not quickly blocked by thirteen stories of brick and glass. Once you are inside the super-block, you are walled in. Perhaps the planners had in mind a precept of Camillo Sitte, the great Austrian theoriest of romantic planning, who remarked how much more pleasant it is to have a gable or a tower greet the eye at the end of a street than the dull fadeaway of an endless vista, but, lacking any variation in roof line or silhouette, any sudden peek at river or sky, what meets the eye within Stuyvesant Town merely accents the prison-like character of the architecture. Even the pool of quiet green at the center of the development—the happiest part of the whole plan—loses a good part of its repose through the inhuman scale of the architecture.

One must not blame the architects and planners too severely. Once the decision was made to house twenty-four thousand people on a site that should not be made to hold more than six thousand, all the other faults followed almost automatically. For this community of twenty-four thousand inhabitants will not have, within its walls, a school or a library, a church or a community room, a motion-picture theater or an auditorium, a clinic, a lying-in home, or a hospital, or any of the other essential facilities for its population. Because of the variety of function of such buildings, there would have been an opportunity for fresh form, too, and this, in turn, might have had a humanizing effect upon the whole design. Considering all the benefits it might have derived from beginning at scratch, on a site as large as this, Stuyvesant Town is a caricature of urban rebuilding.

I have reserved my most favorable comment to the last, and at that it must be qualified. The standards of interior space throughout the project are generous; even the minimum accommodations are more spacious than is usual in prewar public housing projects of the same order, and the kitchens, which on one side are so short that they are almost triangular in shape, are marvelously convenient just because of this fact. The living rooms, all twelve by eighteen feet, are more than adequate by New York standards. I would be more enthusiastic about the quality of this living space if the eighteen-foot dimension were window frontage instead of depth. And about a hundred and twenty square feet of costly space is wasted in each apartment on a windowless anteroom, called the foyer, plus the unusable dark end of the living room. This may be low-rental "economy," but only in Alice's Looking-Glass World.

                                                            1948

FOURTEEN

The Plight of the Prosperous

For a long time, I have been wanting to say a word about the apartment buildings that have sprung up since the war in the wealthy and fashionable parts of the city, mainly on and near upper Park Avenue and Fifth Avenue. The old palisade of apartment houses on Park Avenue, some of them dating from before 1920, had a certain drab decency, with their large, unbroken façades and uniform roof lines, but their mechanical punctuation of windows was not generous enough to give much light and air to their deep interior spaces. In externals, the new vintage seems considerably better; the window openings are often much wider, and many of the buildings have recessed balconies or jutting bays that add a third dimension— and the emphatic contrast of sunlight and shadow —whose aesthetic possibilities we had almost forgotten. In addition, the penthouses, instead of being the afterthought they originally were (many people don't know that they were first used for servants' quarters), are now more deliberately treated with the intention of utilizing terrace and view, though the form they take is largely governed by the municipal setback regulations and is sometimes entirely too accidental to be called "design."

Unfortunately, architecture is not merely the art of putting up a front, and I don't know how I can deal with these buildings without provoking a class war.

While all over town the New York City Housing Authority has been erecting, for the low-income group, skyscraper apartments that provide light and air and walks and sometimes even patches of grass and forsythia, the quarters for the prosperous are still being put up with positive contempt for the essentials of good housing. Though the municipal projects don't do a very good job of meeting the needs of low-income people who have children, they are certainly superior to anything the more fashionable parts of the East Side can offer, if only because they provide light and air on all four sides of the buildings instead of on just two or three. The one thing that prevents many of the new upper-income apartments from being even worse is that they are temporarily protected, in that they are still flanked by the old-fashioned four-story houses that have managed to survive on our side streets. Once those are gone, the outlook from these flats will be completely bleak—or blank. Altogether, a candid assessment of the new apartments for the upper-income groups puts these tenants definitely in the class of the underprivileged. In housing, as in politics, extremes seem to meet.

Curiously, there is nothing new about the plight of the prosperous. I sometimes wonder what self-hypnosis has led the well-to-do citizens of New York, for the last seventy-five years, to accept the quarters that are offered them with the idea that they are doing well by themselves. Apparently those of them who have chosen to remain in New York instead of migrating to the suburbs have forgotten what a proper domestic environment is. Lest someone think that my notions are fancy ones, let me put down what seem to me the minimum requirements for anyone's living quarters. Whether the structure is a single-family house or a thirty-story building, the first necessity is that every room have light and air. Rooms that are in fairly steady daytime use should be oriented to get the maximum amount of winter sunlight. In this latitude, that means that the major exposure should be a southern one, a fact that Socrates discovered twenty-four hundred years ago. To insure enough light and air, the distance between buildings should increase with their height. Our municipal setback regulations make a hypocritical acknowledgment of this principle, but since they were framed to keep land values high rather than buildings low or widely spaced, they have never come within shooting distance of achieving an ideal. The space between buildings should be dedicated to gardens and lawns, partly for beauty, partly to compensate for our tropical summer heat, partly to purify and sweeten the air. Bedrooms should have cross ventilation, or at least through ventilation, and should never face a street. In a city in which apartments are used more at night than in the daytime, the last requirement is particularly important, for noises insufficient to wake slumberers nevertheless have demonstrably bad effects upon the digestive and the circulatory systems. The specification for bedrooms might be modified if a perfect system of air-conditioning is ever invented; I know of none in existence that can circulate an adequate quantity of fresh air in rooms of moderate size without making them uncomfortable. As it is, a layout that permits natural ventilation and yet does not admit too much outside noise is a minimum requirement.

I have not even suggested, in setting forth these specifications, that a family with children should have a house to itself, although every year thousands of husbands move their families to the suburbs and endure all the discomforts of commuting so that they can enjoy the comforts of home. And I have not suggested a private garden for each family, although many people regard this as a basic need for children. After all, even Park Avenue families have offspring. In some respects, such as the provision of pure water and improvements in sanitation, our standards have risen remarkably during the last century, but in the matter of air, sunlight, quiet, and open spaces, we have lost out on every income level, and not least on the highest. Even people who happen to live in urban private dwellings have lowered their standards.

The common row houses, such as those built in the Washington Square district before 1860, met most of these requirements, but the standards have been gradually whittled away, and the richer the occupying family, the more drastically they have been whittled. The addition of a third chamber, between the front and the back bedroom, to serve as dressing room and washroom, was the first step backward; it lessened the possibility of through ventilation. Then the more ostentatious families, with complicated needs for entertainment and service, began to build over their back yards: butler's pantries, dining rooms, extra bedrooms. By the beginning of the twentieth century, they were putting up elegant and costly houses that covered ninety or ninety-five per cent of their land, shutting off all daylight and leaving hardly enough space for an air well between themselves and their rear neighbor. For some reason, the housing regulations introduced to protect the poor against the landlord's rapacity were not applied to the rich to protect them from their own folly, and they were permitted to overcrowd their own living quarters and shut out sunlight, air, and view. If you want a good instance of the results of that folly, take a look at the far from untypical block between Eighty-seventh and Eighty-eighth on Fifth Avenue.

If people of means had been well advised, they would have invested more heavily in land. Thus, they

could have created private parks and squares on the
model of Gramercy Park, and (most important of all)
instead of erecting four- to six-story dwellings on
frontages only twenty or twenty-five feet wide, they
could have built three-story houses on forty- or fifty-
foot frontages—plots that would have provided gardens, sunlight, air, and internal delight. A few score
Manhattanites exercised this good sense, notably the
owners of some of the now demolished mansions on
Riverside Drive, but I know of only a bare handful of
existing houses on a forty-foot front.

This bit of history helps explain the current situation. For half a century, the owners of private residences have repeatedly demonstrated their innocence
of any housing standards, so why should one expect
the speculative investor responsible for the new apartment houses to have more generous aims? No matter
how cramped his flats, he could be sure that his clients
would never find him out. They had no misgivings
that could not be quieted by the provision of an extra
bathroom. As a result, the great mass of upper-class
New York apartments, except for the few old ones
built only on the perimeter of an entire block, have
occupied every square foot of space that our lax regulations allow. Such exceptions as the Apthorp, on
Broadway between Seventy-eighth and Seventy-ninth,
and the Marguery have apparently not stirred investors to emulation.

Only by accident do any of these buildings have
sunlight or pleasant views, and only by accident will
those luxuries remain intact; only by accident are
their bedrooms insulated from traffic noises or protected from the summer heat that rises even at mid-
night from the hot pavements; only by accident are
the rooms well shaped or is the dining place anything
but an airless, windowless hole, grandly called by the
builders an alcove or foyer. (I have frequently come

across fashionable apartments whose foyer, which provides the major internal means of circulation, is also the only practicable dining place.)

Even some of our realty men have become disgusted at the overcrowded and badly designed apartment houses that have been built since the Second World War; they know that flashy-looking balconies that darken the interior without tempting the occupants outdoors, and jutting, chamfered bays with windows that open onto similar windows of the next-door building, only fifteen feet away, are not a substitute for good interior plans. But if one traces the chain of responsibility, one finds that the builder is not altogether to blame; banks lend money on the basis of the number of rooms provided, not on the quality of the space or the soundness of the plan. Because of this shortsighted attitude of the investing organizations, nothing of consequence can be done by the architect. This state of mind comes under the theological designation of "invincible ignorance," for in the design of 240 Central Park South, just before the war, Mayer & Whittlesey showed how well a sound plan, less greedy of filling each last square foot of ground area, can pay off.

Perhaps I can point up these generalities with a brief appraisal of a new East Side apartment house, chosen not because it is a bad example but because it represents almost the best that can be done under present conditions. Since its bad characteristics are common to its class, let this building and its architect remain anonymous. The architect is a skillful hand at the building of upper-class apartments, and in planning this particular one he has, on the whole, made excellent use of the possibilities open to him. For one thing, he greatly increased the amount of window space by using horizontal units that often span a room from wall

to wall. Wherever he had a real architectural opportunity, as in the style and placing of the balconies, he
has done the job handsomely. The gray stone walls
are cleanly handled. If one rated the structure on its
face alone, one would have to give it a fairly high
mark.

But the interior of this building, like the interiors of
all the neighboring buildings, has been governed by
only one consideration, maximum coverage of the land,
and some queer things result. For one, half the apartments on the lower floors have dining alcoves with
neither ventilation nor view, which would be a poor
enough accommodation on a steamship, and even in
the upper reaches two dining alcoves out of four on
each floor suffer the same lack. As usual, the length
of the living rooms is disproportionate to the width—
in this case twenty-five by thirteen feet. That is certainly not an ideal shape, especially when, as happens
here, the only windows are in the narrow wall. Worst
of all, because of a lack of control over the neighboring lots, the southern façade, the best exposure, has
the smallest number of windows in it. All on this side
belong to the kitchens and the upper dining alcoves.
On most of the floors, only two bedrooms out of seven
face the quiet interior of the block.

This is about the best that money can buy in Manhattan today, and the result falls so deplorably short
of decent living quarters that it seems obvious to me
that something more than money is needed if there is
to be a real change for the better in the housing of
the upper-income groups. No speculator assembling
a few costly parcels of land, no bank advancing funds
on the assumption that congestion is a guarantee of
safety for its investment, no builder doing just one
apartment house at a time, no architect designing it
for him can effect any large improvement in this situation. If the upper-income groups want to do better by

themselves, they will have to find a means of assembling large parcels of land and of designing their homes not as single buildings but as units in a large-scale estate, comparable in size and in freedom from through traffic to those erected by the Housing Authority. They would not automatically get good apartments by doing this, for they would still have to find planners and architects who have not slipped into the fatal New York rut, but they would at least have a chance of getting apartments that would compare in the elemental decencies with those now provided by public subsidy for the lowest income groups.

1950

FIFTEEN

The Gentle Art of Overcrowding

For the last dozen years, the New York City Housing
Authority has been carrying out a vast program of
slum clearance and rehousing: a program that has
been going on in many other big cities, it has done
more in that time to improve the living quarters of the
lowest-income groups than all the earlier housing reformers did in a hundred. Acres of dark, musty, ver-
minous, overcrowded tenements have been replaced
by clean, well-lighted, sanitary quarters—also overcrowded. Well then, someone may say, if the Housing
Authority has achieved all this, why ask for Heaven,
too? The answer is simple. If these new housing estates, consisting of tall buildings that hold three hundred to four hundred and fifty people an acre, were
merely a temporary solution of the housing shortage,
they would be admirable. One would not then inquire
why they are dull and bleak, why they are planned
without regard for the human scale and with so little
respect for the needs of family and neighborhood
life, or why the city has devoted itself to a plan of
systematic overcrowding at a time when a lessening
of congestion has become a necessity for urban survival. But the fact is that it will take forty to sixty
years to amortize these congested buildings, and they
are constructed soundly enough to last at least a century, so while the good they do is temporary, the damage they do to the city may be permanent. In Manhat-

tan, where the average height of buildings was, as late as 1931 (when the fifteen-year doldrums in construction began), less than five stories, the Housing Authority has produced in its projects an average height of more than twice that number of stories.

Both the Housing Authority and the City Planning Commission have taken a shortsighted view of this situation. They have treated the shortage of lower-income housing as a disease that can be cured by segregating the sufferers in an isolation ward, and they have overlooked the fact that their favorite type of high-density building has by now created a new pattern of municipal congestion more widespread than the original slum pattern. The Housing Authority, confusing low first costs with low final costs and sacrificing quality to quantity, has planted high-density buildings even in the less crowded parts of the Bronx, Brooklyn, and Queens, and private investors have followed suit, possibly feeling that they now have official sanction for producing congestion. The result has been to reduce the already meager percentage of space for street traffic, for parks and playgrounds, for schools, and for other necessities of city life: but such a grave result seems to have given no concern to the people who have committed the municipality to this policy.

This is nothing new: planned congestion has been practiced in Manhattan for over a century. To understand the situation, one must know something about the conditions housing reformers and housing administrators have been up against. Not least, one must know the long, grim history of slum building and housing reform in this city, which is one more illustration of A.E.'s saw that a man becomes the image of the thing he hates, for every effort so far to wipe out slums has only reproduced the slums in a new form. In creating pockets of deliberate congestion and prefabricated blight, the Housing Authority has been blindly carrying on an old tradition. In the eighteen-forties, in the midst of one of New York's recurrent cholera epidemics, the health commissioner, an able physician named Griscom, pointed out to his alarmed fellow-citizens that the highest incidence of the plague always occurred in the miserable, insanitary quarters of the poor—chiefly in the new "tenant houses," as the first tenements were called. (The first building deliberately designed as a crowded tenement was erected on Water Street in 1835.) That diagnosis led to philanthropic efforts to improve the lowest-income housing, but they were not successful. The first "model tenement," built in 1855, contained so many unlighted and unventilated rooms that it attracted as occupants those who proverbially walk in darkness— thieves and prostitutes. And a Model Tenement competition, held in 1879, brought forth only a new type of blight, the "dumbbell" tenement, so called because of its ground plan, which provided two narrow side courts at the middle of the site. The windows on these courts looked bleakly on the windows, only a few feet away, of the courts of the next-door dumb-bell tenements. The apartments in these buildings were called "railroad flats," because one room formed the corridor to the next. These flats had little light and air and no privacy. Even the philanthropic Alfred T. White, the boldest of the reformers, whose remarkably good tenements, built in the eighteen-seventies and eighties, on Brooklyn Heights, are still standing, departed from the excellent Brooklyn convention of low dwellings by building six-story walkups, creating a pattern of density for Brooklyn that had no precedent even in the poorest parts of that borough.

In the course of seventy-five years of housing reform, no one was able to discover how good housing for the lowest-income groups could be provided as

profitably as bad housing. Slum dwellings earned
twice the rate of return earned by houses designed
for middle-class occupancy in the more respectable
portions of the city, and even philanthropy expected
five or six per cent on its investments in "model dwellings." Thus the slums of New York exuberantly multiplied. The profits were too attractive to the enter-
priser to suggest any sounder mode of building, and
the city had to shoulder the resulting social deficits.
At last, the housing reformers succeeded in persuading the city to enact the "model" Tenement House
Law of 1901, which required better sanitary facilities,
fireproofing of halls and stairways, and windows in
all rooms. Unfortunately, regulations do not create
buildings; unfortunately, too, even the modest demands of the new law raised the cost of new housing.
The poor went on living in their old tenements, and
the old tenements went on deteriorating year by year.
(Those who read Charles Abrams's recent series of
articles on housing in the New York *Post* must have
rubbed their eyes when he reported that windowless
Harlem cellars of the most noisome sort were still inhabited by human beings. But then, hundreds of
thousands of us who pay much higher rentals are
still living in quarters that are definite denials of the
American Dream and even of what we like to think
is the American Way of Life.)

Finally, following the investigations of the state
Commission of Housing and Regional Planning, the
public authorities decided to take a hand in a situation
in which private capital was inert and impotent. First,
in 1927, New York State set up its Division of Housing, which in 1939 began to give financial aid for
slum clearance and rehousing, since it was plain that
only with public loans and subsidies from the state or
the federal government could decent housing be provided for the lowest-income groups at rents they could

pay. But not until the enactment of the federal Housing Act of 1949, which subsidized the purchase of expensive sites acquired through slum clearance, was there any attempt to make sensible building densities possible in metropolitan housing estates. Thus, when the New York City Housing Authority was set up, in 1934, it soon succumbed to the temptation, in the interest of lower first costs, of maintaining in its projects the population density of the slums that had been razed to make room for them. To correct a pathological condition, it employed a pathological remedy. Most of the ills that have followed resulted from that yielding to temptation, which froze the design of the projects in an obsolete, congested mold that ignored the needs of the tenants and the welfare of the city. The Housing Authority has even seemed to pride itself on the rigid formula under which it has set its architects to work, perhaps because it mistook the tall, free-standing buildings of its projects for Le Corbusier's meretricious City of the Future. In all its career, the Authority has not succeeded in effecting a single significant economy in these standardized dormitories, because significant economies—such as the ingenious ones in construction and plan the architectural firm of Holsman, Holsman, Klekamp & Taylor has introduced in certain Chicago apartment houses—cannot be achieved in terms of tall and necessarily expensive mechanized apartment houses. Such picayune "economies" as the Housing Authority has introduced—like omitting doors for the closets in apartments—only throw a burden on the tenants.

Against the physical improvements in the flats provided by the new housing, one must set the massive damage high-density housing has done, and threatens to do, to the city. If New York continues to be made over in this fashion, the ensuing congestion will throttle it, and the weekend exodus from its discomfort will become even more harrowing in direct proportion as it becomes more compulsive. Temporary and futile palliatives of the kind so avidly proposed by Mr. Robert Moses (more superhighways and tunnels to pump more people into the suffocating center of the city) will absorb funds badly needed for schools, hospitals, homes for the aged, libraries, and other facilities already in stringently short supply. One example of the cost of congestion will stand for a score of others. A while ago, Mr. Douglas Haskell, of the *Architectural Forum,* figured that the city could, even at present land values, save a great deal of money by building one-story schools of a new and flexible design. But the school authorities, while admitting the desirability and the probable economy of such buildings, plaintively replied that "the erection of these great building-canyon areas, to densities of four hundred to six hundred people to the acre," in the new housing projects would force them to "fill the Hudson River" to get the requisite space.

To revert to an earlier point, such economies as the Housing Authority has effected by its intensified occupancy of the land are only first-cost economies. The final effect upon the municipal budget is disastrous. Moreover, the congestion within the housing projects increases their grim, depressing atmosphere of social segregation. Though the Housing Authority, with admirable zeal, has abolished racial segregation, the law under which it operates has established segregation by income, a class-distinction concept all too plainly embodied in the Authority's great building units. Again, under the terms of the law, this housing has been separated from every other normal manifestation of neighborhood life, so although these estates are physically neighborhoods, or even small cities, they lack most of the organs and attributes of a full-fledged domestic community. The fourteen thousand people who live on the thirty-four acres occupied by the adjoining Jacob Riis and Lillian Wald Houses, in lower East Side Manhattan, haven't a single church or synagogue or motion-picture house or public market within their whole area. The provision of these facilities, in low, freestanding buildings, would do more to compensate for the bleak uniformity of the huge black prisms the tenants occupy than any amount of fancy ornament. There is no architectural substitute for the variety and stir and color of a real neighborhood.

This situation can be relieved in one way: Income groups should not be segregated by neighborhoods, and the Authority should be given the power to provide public markets and shopping centers, as the New York Life Insurance Company did for its Fresh Meadows housing development in Queens, and sites for other public buildings. Enterprisers like the founders of Levittown, that enormous residential project on Long Island, have discovered that the main profits of such a venture are derived not from the sale of land and buildings but from the rent of the shopping centers.

Each year, thousands of taxpayers are driven to the suburbs because a healthy pattern of life does not exist within the heart of the city. That is poor planning and poor business. We must abandon the old habit of giving housing the isolation-ward treatment. We must set sane limits of density in both business and residential parts of the city, and we must decentralize many commercial and manufacturing activities that are now overconcentrated, to no one's real benefit, in Manhattan. The initiative, as I have said before, should come from the City Planning Commission. As for the Housing Authority, it is time that it retraced its steps to Harlem River Houses (in upper Manhattan) and Williamsburg Houses (in Brooklyn) both built a dozen years ago, and urged the city to use the new federal Housing Act's beneficences to establish lower densities, on a pattern that would create an open, healthy, and beautiful city. What has been achieved in Fresh Meadows is a reasonable standard for new urban housing and city building everywhere. Should we go on rebuilding New York on the obsolete patterns of Parkchester, Stuyvesant Town, and the municipal housing projects, we should merely be exchanging slums for superslums.

1950

SIXTEEN

Schools for Human Beings

Since the war, the Board of Education has been putting up a series of new schools all over the city, some
of them planned as early as 1943. They are among
the happiest signs of architectural decency and civic
sanity the metropolis can show, for they are almost
the only new buildings that seem aware of the human
scale, and they recognize the importance of that scale
in the many activities a school now harbors. This is
all the more remarkable because the New York City
Housing Authority has been heedlessly galloping in
the opposite direction, and because the Board of Education itself, when, in 1939, it built the Joan of Arc
Junior High School, on West Ninety-second Street,
seemed about to compound the congestion of New
York's huge skyscraper housing projects by creating
equally congested skyscraper schools. These new
ones have sprung up in many parts of the city—from
Fresh Meadows to Harlem, West End Avenue, and
South Brooklyn—and certain features characterize
all of them. They are freestanding, no more than three
stories high, set back from the building line on three
sides, and usually protected, as far as light and air are
concerned, by a playground. Sometimes, indeed, a
school and its playground occupy a whole block.
None of these schools is quite as handsome as the
best ones in other cities—those designed by Kump
or Neutra in California, by Saarinen or Perkins in

the Middle West, and by equally competent school
architects working in smaller towns nearer home. But
the elements of good planning and sound design are
obvious in them, and it will need only a little further
exercise of the educational administrator's imagination
—and a little extra money for the aesthetic imponder-
ables—to produce schools as good as they can possibly be.

Since these new schools are all pretty much of a
piece, I shall discuss in detail what is architecturally
one of the best of them. This is Public School No. 33,
just south of Chelsea Park, which is at the north-
eastern corner of the superblock bounded by Ninth
and Tenth Avenues and Twenty-sixth and Twenty-
eighth Streets. I was attracted to the neighborhood
because it is in itself a sort of architectural museum.
A block north of the school, and across Ninth Avenue, rises the baroque green copper spire of the
Church of the Holy Apostles. Straight to the north, set
in the park, is the funereally decorated Health Center, done in seedy undertaker's modern, and west of
that looms the grim and toothless ghost of the original
P.S. No. 33, a now abandoned four-story building of
grimy buff brick and classic brownstone trimmings,
with a Dutch gable on one side and a formidable wall
surrounding the even grimmer extension to the west.
Nearby stands a public bath and gymnasium of the
gaudy terra-cotta period, circa 1910. West of this
whole architectural hodgepodge rise the hulking structures of Elliott Houses, one of the most overcrowded
and dismal examples of public housing in the city—
though, I hasten to add, a paradise compared to the
evil tenements they replaced.

The park and the half-block allotted to the new
school are now in a sense a unit, for the city has
closed Twenty-seventh Street from Ninth Avenue
halfway to Tenth, and installed a walk that extends

all the way to Tenth beside the surviving western segment of Twenty-seventh Street. The possibilities offered by this site have been exploited with good effect
by the architect, Mr. Eric Kebbon. The school, which
fills only a part of its lot, has a ground plan roughly
in the form of a Saint Andrew's cross. The northeast
and southeast arms of this cross abut on Ninth Avenue, and the main entrance opens on the triangular
courtyard between these arms. The end wall of the
northeast arm bears, in bold aluminum lettering, the
name of the school, flanked by equally bold limestone
seals of the Board of Education and the City of New
York. To emphasize the public nature of the structure, the stairwell to the right of the main entrance is
in a tower embellished by a clock whose hands and
numerals are also of aluminum. The northeast wing
is two stories high, the other wings are three stories,
and the clock tower is four. The stairwells at the ends
of two of the wings likewise jut above the main roof-
line. As a result, there is a continuous dynamic movement of the planes of the building as one walks around
it—and, miraculously, one *can* walk all the way
around it—that is entirely lacking in buildings conceived as uniform slabs. This dynamic quality has
not been exploited to the full in this case, but at least
a beginning has been made. (How effective an aesthetic resource this is was amply demonstrated by the
recent exhibition of Frank Lloyd Wright's model of the
S. C. Johnson & Son research building at the Museum
of Modern Art.) Like most of the new schools, P.S.
No. 33 was designed to house from twelve to fourteen
hundred pupils, which is about what the older vintage held. But because the new buildings are only
three stories high and are set back from the building
line, they look much more genial and intimate than
the high, crowded buildings of fifty years ago.

The lively quality of the plan and elevation of the

Chelsea school is, unhappily, weakened by some of the details. The stairwells are lighted by a great vertical panel of glass brick above the entrances. Since the brick façades are severe to the point of bleakness, glass panes, in solid colors or in geometrical patterns, could have been used to advantage here. The deep-yellow glass that the Dutch often employed in their old churches creates a wonderfully golden interior; such glass, too, would have given the windows of the auditorium a rich contrast to the magenta curtains that drape them. Furthermore, a livelier color scheme could have been used for the window trim and the doors, whose dull buff or green too easily succumbs to the inescapable grime of the city. The instincts of the school children are here superior to the current habits of architects; the cutout figures and the paintings done directly on glass that the pupils have used to decorate the lowest panes of their classroom windows give these façades an animation that is otherwise lacking. Here the client has something to teach the architect who is willing to listen.

But it is in the interior that the greatest architectural innovations have been made in the city's new schools, and to realize how great these are one should think back to the sort of schoolhouse that was "modern" forty or fifty years ago. Squeezed between tenements on either side, it was usually five stories high, done mostly in limestone, and terminated in turrets that had, like the main entrance, vestigial Gothic traits. The ground floor was a dark space called the "indoor yard," a dismal hall that served only as a play area at recess and as a marshaling area where those who came early to school forgathered to be regimented into ranks by monitors before marching to their classrooms. (Oh, the dreary minutes between gongs!) The main stairs, sacred to visitors and teachers, led to the principal's office, usually on an upper floor. The lavatories and the drinking troughs were mostly in the basement, a throwback to primitive days indeed. There was no auditorium; instead, six classrooms on an upper floor were laid out in a double row and separated from each other only by folding doors, which made them visually but not orally private. For morning assembly, these doors were opened to disclose the principal's platform and the Bible lectern at one end. None of the building was usable for any after-school or adult activity except the indoor yard, and that was generally used only once a year, as a polling place. If there was a gymnasium, it usually occupied part of the top floor, and the heavy thump of calisthenics was all too audible on the floor below. Even when they were new, these schools were dismal and depressing places, and not a little frightening to the very young by reason of their immensity. Economically, they were a combination of wastefulness and makeshift; architecturally and educationally, they completely lacked the human touch.

The new schools are the result of a remarkable change in educational and municipal policy during the last half-century, and particularly in New York during the last twenty years. The first innovation was the provision of a real auditorium and a spacious, ground floor gymnasium. Then came the convenient transferal of the clothes closet from a special room or a cage in the corridor to the classrooms. This was followed by the school cafeteria, which was started during the depression of the nineteen-thirties to provide milk or school lunches for undernourished children. Then came various rooms to take care of new services and specialized activities—now a "cardiac room" with reclining steamer chairs for students with bad hearts; now a room for children with defective eyesight, with illumination whose intensity is maintained by an automatic device that operates when even a passing cloud decreases the light. But there are still some members of the school community who have been overlooked. No school that I have yet seen has made any provision for housing comfortably the volunteer guardians (usually parents of the children) who stand at the entrances to check on people who wish to enter the building and guard against the ever-present threat of vandalism. They have neither a sentry box nor a concierge's window to protect them in cold weather.

Even more important, of course, is the change in educational policy, leading to increasing emphasis upon the pupil's participation and upon constructive tasks in art and science, and less attention to verbalisms and abstractions—doubtless, in some cases, too little attention. One of the most marked changes is in the relationship between the young and their teachers; there is much less formal discipline and more affection on both sides. A good part of all this is reflected in the architecture of the new schools. It is not yet fully reflected, though, because educators, building administrators, and architects are still not always able to speak the same language.

Beyond the main entrance of the new Chelsea school is a wide lobby, paved in terrazzo, its walls covered from floor to ceiling by pleasant large green tiles. This generous hall is separated from the narrow ones that give access to the classrooms by fire doors, which also serve to shut off the rest of the building at night, when the auditorium and the gymnasium, opening on this hall, are used for adult activities. The gymnasium is a spacious, well-proportioned room, with windows set high in the walls, a sound-absorbing ceiling of fiber blocks, and tiled walls in a warm tan. Up-to-date showers and dressing rooms are provided. A nurse's room for physical examinations, and even an isolation room, with a cot, for children who come down with a serious illness during the day, are also provided. The auditorium, which seats four hundred, has not merely a fully equipped stage but dressing rooms in the basement. This equipment might seem lavish if the school's daytime needs alone were considered, but a basic principle of these new buildings is that they shall be available for adult use at night, and not simply for balloting or registration for ration cards, either. To this same end, the teachers' lunchroom can be used as a meeting place for neighborhood committees and groups. In short, the new schoolhouses are neighborhood centers as well as educational buildings, and they are designed to house a wide gamut of social activities, with a minimum of expense. Another characteristic of the new order of things is that the administration office and the principal's office, both far busier places than they used to be, are on the ground floor. The principal is no longer a dim, distant, forbidding, frowning deity, and the growing volume of human traffic his office handles, as his human relations and scholastic duties multiply, is thus expedited.

To get space for all these new activities without extra expense, the architects of the latest batch of schools have eliminated the indoor yard. That gain is no small one, even though (particularly in inclement weather) it raises the problem of what is to be done with children who get to school before the teachers have opened the classrooms. The provisional answer most principals have been driven to devise consists of asking the parents to hold on to their young ones till the last possible moment, and to put all early birds in the auditorium until the teachers arrive. A better solution than this must be worked out. The introduction of the cafeteria creates another traffic problem. If an attempt were made to serve lunches to everyone, there would be hardly enough space for any other activity in the whole building. As it is, P.S. No. 33 feeds only a third of its population— four hundred and thirty young souls—in shifts between eleven-thirty and one. The only practical answer to this problem is to provide a multi-purpose room opposite the lunchroom, and even that raises the secondary problem of disposing of all the tables and chairs and putting the room to rights after lunch —a duty small children can hardly perform themselves.

In No. 33's classroom wings, the innovations are no less conspicuous. The low ceilings of the corridors and stairwells are covered with acoustic tile, and the walls are lined halfway up with glazed gray tiles. In first cost, these tiled walls may be expensive, but in final cost, considering the expense of cleaning and painting plaster walls, they are an economical as well as a pleasant choice. The classrooms may seem narrower than the old-fashioned kind, but that is because they are actually longer. Clothing cupboards, bookshelves, and exhibition panels line the wall opposite the windows, while in the rear there is a drinking fountain, as well as a sink with running water—a hitherto unknown luxury now indispensable in teaching art and science. Until recently, most classrooms had benches and desks lined up in rows and anchored to the floor, one of the Dickensian Gradgrind and McChoakumchild's early inventions. In the new schools, the tables and chairs (two children to a table), made of wood and tubular steel, are movable, and the teachers are encouraged to rearrange the classroom to fit the requirements of the lesson. (Some of the older teachers, poor dears, are still a little uncertain about how to use the new freedom.) In the painting of the interiors, old habits, unfortunately, still prevail. Once one has abandoned the notion that buff is the sole respectable color for classroom walls, there is no good reason to fasten on a single shade of green. Indeed, there are many good reasons for using more than one color.

I have left to the end one of the most charming features of the Chelsea school—the provision of glass-paneled showcases in every corridor for exhibiting the children's paintings, sculptures, and craftwork. These cases give the corridors a liveliness they would otherwise lack. The idea could be carried even further, and the main lobby, with its ample wall space, could become an exhibition hall with movable wall screens, which could be devoted to paintings, and perhaps with niches for the exhibition of sculptures and ceramics. The fact is that architects have not yet caught up with the lessons of sound modern education, whose emphasis is on experiment and creative effort. There is nothing so deadly to the imagination of youth as a finished and perfect building. And if the architect leaves more room for art, there may be less temptation to vandalism. On this I intend to say more at a later date. Not the least merit of Mr. Kebbon's design for the Chelsea school is the hope it gives for something even better in the future.

1952

SEVENTEEN

Municipal Functions and Civic Art

The original core of almost every American city is by
now a clotted mass of antiquated buildings. Sensibly,
without waiting for bombs to clear the ground, many
cities are already at work on extensive demolitions and
planning equally extensive reconstructions. St. Louis's
seedy waterfront area is about to become Jefferson
Memorial Park, "the gateway to the West," symbolized
by Eero Saarinen's great parabolic arch. Pittsburgh's
Golden Triangle has been cleared to make room for a
group of gigantic skyscrapers, unfortunately conceived
in the image of Le Corbusier's now outdated Voisin
plan for Paris. Philadelphia has razed its Broad Street
Station and the obstructive Chinese Wall that carried
the tracks of the Pennsylvania Railroad to it, and even
more elaborate rehabilitation is going on around Independence Hall, on land acquired by the federal government with the aim of preserving the whole outdoor
museum of historic buildings that surrounds it and of
providing an ampler setting for the principal monuments. But where does New York stand? Well, New
York—or, specifically, Brooklyn—has started one of
the most extensive of all the programs of demolition
and rebuilding. To that end, the municipality has acquired by condemnation and purchase a vast downtown area, bounded roughly by Fulton Street, Jay
Street, and Sands Street. And already, though part
of the site of what is to be a new civic center is still

covered by grimy buildings waiting for vandalism or fire or the wrecker's crowbar to level them, a large portion near the Brooklyn Bridge has become a public park. Now, while the reconstruction is still in its beginning stage, is the time to take a look at this promising sweep of space, extending from Borough Hall to the Brooklyn Bridge. Some of the program evokes one's admiration, some of it needs revision, but as a whole it deserves support.

People who have had any business to do in the Borough Hall area have been aware for a decade of the changes that have been going on in this dilapidated district. They began with the tearing down of the old Fulton Street "L" and the elimination of the vast wilderness of trolley cars that circled Borough Hall or went over the Brooklyn Bridge. Long ago, when Mr. Cleveland Rodgers, once a City Planning Commissioner, was editor of the Brooklyn *Eagle,* he started a campaign for the improvement of this sordid area, and he has lived to see not just the implementing of a plan that should fulfill his early dreams but a vast expansion that includes the whole district from Atlantic Avenue to Wallabout Bay and from Fort Greene Park over to Brooklyn Heights, along which stretches a great promenade fronting the East River. This is the biggest piece of redevelopment the municipality has ever attempted, on a far vaster scale than any of its housing developments, and it requires a much more complex job of planning. If this work is well done, it will provide a positive example for the reconstruction and modernization of the rest of the city.

City planning of this order is not achieved on the drafting board alone. Still, the part played by imaginative design must not be underestimated, nor should the depressing effects of mediocre design, also visible in this effort, be too complaisantly accepted. To rebuild this area demands the cooperation of over fifty public and semipublic agencies, and it involves every kind of municipal enterprise, from highway building to public housing, from public buildings to the rerouting of traffic. Each of these agencies has its own needs, its own purposes to consider, sometimes its own vested interests or entrenched prejudices to protect. To emerge from this welter with a plan that does justice to each organization's needs and also serves the long-time interests of the city requires political finesse of the highest order. Perhaps the most powerful agent of persuasion is a large guiding idea, like that which Daniel Burnham brought half a century ago to the improvement of the Chicago lake front. The changes to be made by the Brooklyn reconstruction are far-reaching. For instance, when the road building is complete, it will be possible for a resident of downtown Brooklyn, once he has reached the great arterial Belt Parkway, not to encounter another traffic light for the next two hundred miles if he heads in the right direction.

The problem of replanning this huge area is threefold. First is the determination of the right functions for the various portions of it. This is of vital importance now, while there is still time to make changes. For it is by no means certain that, as is intended, this site should be dedicated to public buildings, housing, and open spaces alone. The second task is the recasting of streets, avenues, promenades, and gardens so that the bulk of wheeled traffic will be routed around it rather than through it and the open spaces will add charm and dignity to every part. The third job, the hardest of all when so many agencies must cooperate, is grouping the buildings and limiting their height and density of occupation so as to create a handsome and durable architectural framework that will serve Brooklyn as many centuries as the buildings around the Place Vendôme have already served Paris. Such an effort requires both audacity and patience. Recently, a federal courthouse was projected for nearby Brooklyn Heights, which should remain a residential district; a sounder program would place this building in the new civic center. Again, the old Post Office, a decent example of its period, could well be retained within the area, but the Post Office Department wants a new one on a site it considers more suitable for its traffic needs. The old Brooklyn Public Library, which is outside the area, on Montague Street, has declined the offer of a site within the area, and it is probably right in doing so, since it now performs only a neighborhood function. There will presumably remain for some time in the new civic center such loft buildings as the one that houses the State University of New York's Institute of Applied Arts and Sciences, solidly built but an eyesore of raw concrete and patterned brick. And there is also a premature housing development, Concord Village, which it will be hard to merge into a general scheme for the area.

In determining that this part of the city should be used for public and civic buildings, which are now crowded on one side of Borough Hall Plaza, the planners made a sound decision. The main business district of Brooklyn should ultimately in large part be decentralized in other sections of the borough; the Erie Basin area, the Fort Hamilton area, the Flatbush area, the Greenpoint area would all profit by large-scale replanning for business use, and the creation of other focal points for business would greatly increase the value of the present transportation facilities by making the rush-hour traffic not merely a one-way proposition. The partial dispersal of department stores from the downtown area is another important consideration. But at the same time, part of this area could be the scene of a bold municipal experiment in planning a whole business precinct of offices and shops. Despite the example of Lever House, I doubt that business has yet acquired the civic intelligence to pioneer in this kind of large-scale planning without introducing, for immediate profit, the congestion that must be avoided. Now that the city owns this land, it could lay it out in a new, low-density pattern and invite investors who are interested in stability and security, such as life-insurance companies, to do the actual building on long leases.

Luckily, a new structure in this area, finished last year, has already set the right precedent for a modern office building, and I am hopeful that its sobriety and efficiency will leave a mark on the whole neighborhood. It is the Board of Transportation's new offices, a building that is well conceived, straightforward in design, well lighted, practical, and businesslike. (The original architect was William E. Haugaard, who was succeeded, upon his death, by Andrew J. Thomas.) This large structure fronts on Jay Street, and two wings, one higher and one lower than the main part, run back toward Pearl Street. The main section is thirteen stories high. The façade throughout is composed of smooth limestone, and there is a neat Monel-metal cornice to punctuate the top. The windows are large panels of glass set flush with the wall. Each panel consists of one large fixed central unit, with two openable casement windows at each side and horizontal units above and below, which are also openable.

This façade is simplicity itself. Though the building does not employ the cantilever construction that would have permitted continuous banks of windows, uninhibited by the presence of supporting columns in the outer walls, there is only an almost imperceptible diminution in the amount of light that enters. The main part, though it is three hundred and fifty-six feet long, is only eighty-two feet wide, and what is even more important is that it is only thirteen stories high. These are ideal dimensions for a building that is intended to rely upon natural lighting and natural ventilation; above all, they are the ideal dimensions for a business district that seeks to eliminate the handicaps of congestion. As it is, some twenty-five hundred people occupy this building, and that is quite enough.

In short, this seems to be the very model of an efficient office building. Not a cathedral of commerce, not a temple of advertising, not a palace of municipal power: just a group of offices arranged for the efficient dispatch of administration. That straightforwardness, that underemphasis come like the firm voice of Elmer Davis after a perfervid commercial. For all that, the architects did not overlook the fact that this simplicity makes every departure that much more striking. For one thing, the three units of the building, just because they are of different heights and axes, are in dynamic relation to the approaching spectator. Another happy touch is that the architects, with great imagination, chose to reveal the fire exits, instead of trying to conceal them, through three great square openings on each floor. The aesthetic effect of this treatment is as powerful as it is simple. With the aid of light and shade alone, the architects turned what might have been a blank façade into an expressive one.

In the interior of the Transportation Building, straightforwardness is carried to the point of austerity. There is shop and office space on the street level; parallel to the street runs a long corridor, and at each street corner of the main building an opening takes subway passengers quickly off the street and down to a commodious underground concourse to the Jay Street station. This is another example of spacious monumentality, without a single false touch; even the standards that indicate the subway entrances are in good taste. Because of the architects' reliance on light-gray granite for the exterior of the entrances and on

travertine for the interior passages, with only white
bronze as a counterfoil, the total effect is perhaps a
little cold, especially when the corridors are empty.
And unfortunately the pastel colors of the walls and
the trim in the offices are likewise needlessly bleak.
But there is nothing in either the mass or the detail
that will look antiquated or comic fifty years from now.
The architects of the Transportation Building plainly
meant business, as surely as Burnham and Root meant
business when they designed their classic Monadnock
Building in Chicago. The reward for that continence
is a building that might easily become a model for a
whole district of structures planned for efficient work,
for easy access by public transportation, and for convenient pedestrian circulation.

One must also say a special word of commenda-
tion for the terminal feature of the long ground-floor
corridor—a gray granite wall, which even from a distance discloses that a map of the world is spread
there, in polished (and so darker) granite. When one
gets nearer, one discovers that this is a roll of honor
for the employees of the Board of Transportation who
served in the Second World War. Every theater in which
they served is designated, and their names—and a
gold star, if that was their portion—are printed where
they belong on the map. The whole message is deliv-
ered with grave simplicity, yet it is one of the most
stirring war memorials I have seen, in the way it brings
home the facts of a global war and the part that these
men played in fighting it.

A business district that consisted mainly of buildings of this type would be far more efficient and far
more handsome than anything the rest of the city could
show. It would have charms that, if less striking than
those of our cloud-capped towers, would give greater
promise of durability. I hope that the city-planning
authorities will accept nothing in this redevelopment

that does not carry on this good example. We could cheerfully trade the imitative classic monumentality of the old civic centers, with their reliance on columns and cornices, and the superficial modernity of many recent skyscraper designs for some of this honesty, this straightforwardness, this matter-of-fact decency. Properly grouped, a half-dozen Transportation Buildings would create a new kind of urban space, distinguished by composure and order.

1953

EIGHTEEN

Closed Minds and Open Spaces

For the reconstruction, now in progress, of the Borough Hall-Brooklyn Bridge sector of Brooklyn, which is to become that borough's new civic center, the city acquired a vast area largely occupied by dingy buildings that had long begged for demolition. Borough Hall Plaza, an irregular polygon, lies at the south end of this area; the entrance to the Bridge lies at the north end. Fulton Street sweeps around the western side of the area in a curve; Sands and Jay Streets are the other boundaries. The origin of this large-scale improvement dates back to the W.P.A. days of the 'thirties, when S. Parkes Cadman Plaza, as the lampposts call it, was begun. This is at the north end of the area. Just south of it is a pleasantly conceived and much more recent sunken garden. Unfortunately, this garden has more than lived up to its name, for it was placed on land that had not been soundly filled, and it will have to be regraded and repaved. Before that happens, I trust that the whole treatment of this civic center will be reviewed —and revised—by all the authorities concerned.

Cadman Plaza is a long stretch of formal garden with a grass plot in the middle and a big limestone wall, which turns out to be the rear wall of a war memorial, in the middle distance, flanked in the foreground by a comfort station and a park-maintenance building. East of the plaza is the new private housing development called Concord Village. The block be-

tween them will eventually hold a Red Cross headquarters and a federal building, both surrounded by
open space. For the first time in the seventy years of
its life, the Brooklyn Bridge will have a decent pedestrian approach, and since it is the one bridge across
the river designed for the delight of pedestrians, this
is no mean improvement in itself. The plaza is far
from finished, but the outlines are already well established. The task of coordinating all the municipal
activities and functions that are to be housed in this
area is a huge one, and some of the basic problems in
design have not been solved; indeed, as yet they have
hardly been considered. (Incidentally, what has been
worked out up to now can be inspected in a large
model in the entrance hall of the Borough Hall.)

Both Manhattan and Brooklyn are poor in open
space that is accessible to their more densely settled
areas. When I lived on Brooklyn Heights, I had to take
a subway to Prospect Park if I wanted to stretch my
legs in the sun within sight of greensward. So one is
tempted to welcome without reserve any additions to
such space. And when this particular addition comes
as a result of tearing down shabby, obsolescent buildings, the gift seems so enormous that one hardly feels
like asking what use is to be made of it. Yet the very
rareness of this opportunity also brings to light something else, which one first became uncomfortably conscious of in the site plan for the United Nations—the
fact that New York architects and city planners are so
unused to handling open spaces that when they are
offered the chance, they don't know what to do with
it. The simplest way of treating Cadman Plaza was to
turn it all into a formal park, and that is what has been
done. But what will be the aesthetic effect of this treatment, even after the multiple rows of plane trees and
the defining yew hedges rival those in Bryant Park?
Precisely because of the sweep of space that has been

opened up, what one sees from it is the horizon of industrial Brooklyn—a jagged sky line of warehouses
and factories not notably embellished by the new fifteen-story buildings of Concord Village. Indeed, these
gawky buildings sound the wrong note for the whole
development, and the situation will not be improved
by setting more housing projects, on the same pattern,
in other parts of this area, as is now planned. For the
sake of a long vista, one might temporarily accept a
narrow segment of that grim sky line, but the ragged
fringe of the old buildings that hem in this plaza is an
aesthetic disturbance, and the removal of it is a major
problem. Merely to produce an open space and hope
that the ugly buildings surrounding it will be replaced
by good ones is not effective design.

Do not think I would have the city forfeit a single
square foot of ground now dedicated to parks, gardens,
and plazas. But there is a better treatment than a
wholesale concentration of open space. A great public
square in the civic center of Brooklyn is certainly in
order, but it should be planned for parades and public
processions, for ceremonies and celebrations—not as
a neighborhood park. Neither the sunken garden nor
the plaza, with its long stretch of verdure, suggests
that any such public use was in mind. The plaza will
be a pleasant place for nursemaids and mothers with
baby carriages, and for public employees during a
spring noon hour, but that will be about all. Here is
a place that obviously should serve as a great meeting
place for crowds—a "core" in the sense that the architects of the Congrès International d'Architecture
Moderne have given the word. (See their book, *The
Heart of the City*, recently published here.) Nothing
in the design permits or even suggests such a use. The
planners might have remembered with profit the old
approach to St. Peter's, in Rome, and have planned,
midway between Washington and Fulton Streets, a

broad processional way, flanked with trees and public buildings, that would open suddenly, like the great colonnaded plaza in front of St. Peter's, to receive the crowds that would pour into the area on a public occasion. That would have the advantage of concentrating the new public buildings and, in the very act of doing so, not just adding to their convenience but using them as a frame for public ceremonies. Characteristically, alas, the only procession the planners sought to take care of was that of motorcars. Widening Adams Street to a hundred and sixty feet, they propose to lead its ten lanes of cars from the Brooklyn Bridge to Borough Hall. Thus they will pierce their new civic precinct with a major traffic artery, in defiance of every canon of sound civic design, and will nullify by this massive disruption most of the aesthetic and social qualities this center might have possessed. As for the resulting congestion at the junction with Fulton Street, that is almost unimaginable.

The fact is that one cannot have good architecture, either functionally or aesthetically, unless buildings and open spaces are conceived together. The right interval is as important as the right note or the right succession of notes. I suspect that one reason for the commonplace and unimaginative and sadly shortsighted utilization of this magnificent opportunity is that park operations are handled in one city department, street planning in another, and building operations in a whole series of other agencies, municipal, state, and federal. Because of that division of labor, everyone involved tends to hold to conventional practices instead of applying a fresh imagination toward the conception of an entirely different urban pattern.

This situation has affected the war memorial in Cadman Plaza, one of those anomalous structures nearly every American city now boasts, which testify to the pious intentions rather than to the understand-

ing of those who have fostered them. It is a meeting
hall, the front of which, facing Borough Hall, is a
simple wall of limestone bearing an inscription and
flanked by two gigantic limestone figures. This building is not very useful, is not very beautiful, and is not
in the least eloquent. In fact, it is principally a warning
of what may happen to this entire civic improvement if
the easy and obvious stereotypes of an older day prevail.

Let me give one more example, a smaller one, of
the failure to think clear through to a new order of
design. Along Tillary Street, between Adams Street
and Jay Street, bordering Concord Village, runs a
tree-lined promenade. If the present cross streets were
done away with, this promenade would set a pattern
for pedestrian communications all through this new
area, for such green lanes would unite both open
spaces and buildings. Instead, the plans for the widening of Tillary Street will eliminate the entire promenade.

What this new civic center needs is not merely open
space but a more prudent distribution of that space in
relation to the buildings that will occupy the area. Setting aside one huge slab of open space and allowing
the rest of the area to be built up in the usual monumental manner is no solution. Precedents like the
Washington Mall and the Benjamin Franklin Parkway,
in Philadelphia, show how obstructive and self-defeat-
ing generous public spaces can be when their purpose
is misconceived at the start and their actual mode of
operation ignored. This civic center should have a
number of gathering places, human catchment basins,
to bring people together; Cadman Plaza, as it is now
conceived, is an area of dispersal, whose inviolable
green lawns naturally reduce the number of people
who can congregate. Happily, the planners have made
a beginning toward providing the right kind of space

in the little semi-enclosed plaza they have planned next to the Board of Transportation Building, between Jay and Pearl Streets. This space will serve as an approach to the projected Supreme Court Building, and, rightly handled, should be very pleasing.

Besides promenades free from wheeled traffic, running through the interiors of superblocks, one would like to see occasional squares, also completely insulated against wheeled traffic but perhaps linked together by promenades. They might be only half the size of Gramercy Park, but if they happened frequently enough, they would be a boon to people seeking an hour of refreshment in the open. Just outside the new center, at Montague and Henry Streets, there is a wide sidewalk that, because of low buildings across the way, gets the afternoon sun. There, on a winter afternoon, I counted eighteen mothers with their baby carriages, huddled together in the sun and air, even though there was no place to sit down. True, they could have had a whole park to themselves that day if they had gone to the harbor end of Montague Street, but they had better shelter where they were, surrounded by buildings that acted as windbreaks—an important consideration in winter—and most people aren't willing to walk more than a quarter of a mile for an airing when they are tethered to young children. Little parks and gardens should occur at quarter-mile intervals, or thereabouts. Better ten half-acre parks scattered through a neighborhood than one five-acre park too far away from home for too many people.

Fortunately, if Cadman Plaza is the wrong kind of open space for the embellishment of a civic center, the right kind for neighborhood recreation is not far away, inasmuch as the Parks Department—i.e., Mr. Robert Moses—took advantage of the upheaval caused by the construction of the Belt Parkway to

transform the whole area on the waterfront bluff
above Furman Street, from Remsen Street over to
Orange Street. Furman Street runs parallel to the
harbor, and on the bluff above it runs the street called
Columbia Heights, which is lined on both the harborward and landward sides by residences, mostly commodious old houses of another era. Where the cross
streets intersected Columbia Heights, there were little
breaks in the wall of harborward houses, and one could
peer through an iron fence at bits of the harbor and
downtown Manhattan. At the southern end of Columbia Heights, where Montague Street joins it, there was
a paved plaza, opening on the harbor and filled with
benches from which one had a grand view. Now, just
above the viaduct of the Belt Parkway, there stretches
a wide pleasance, lined with benches, planted with
rows of trees, and flanking the back yards, gardens,
and rear façades of the houses on Columbia Heights.
The East River side of this parkway is, for part of the
distance, still lined with the warehouses along Furman
Street—the street that so roused the interest of Ernest
Poole, the author of that minor classic *The Harbor*—
and the park will not achieve its full capacity for recreation till these buildings are leveled or replaced by
lower ones. But even now this promenade gets the full
afternoon sunlight, the salty air of the Bay, and all
sorts of cheerful glimpses of domestic felicity in the
private gardens, and at either end there are untram-
meled and magnificent views of the Bay, lower New
York, and the Brooklyn Bridge. For centuries to come,
people will probably bless Mr. Moses and his colleagues for having had the imagination to take advantage of this opportunity and to carry it through with
such dash. Indeed, our recent waterfront esplanades
—I am thinking, too, of the one on Riverside Drive
from Seventy-second Street to Columbia University,
and the one with the handsome curved railings that

edges Carl Schurz Park, on the East River, both of which can be credited to Mr. Moses—must be counted among the most satisfactory accomplishments in contemporary urban design.

There is one place where the new Brooklyn promenade rises to a breathless architectural climax. At the northern end, it terminates in a small circle surrounded by a stone wall. From one side of this circle, a reverse spiral leads upward in a ramp to the street. Within the circle, there are benches and trees, all arranged in the same circular pattern, while the stone paving consists of circular bands of contrasting textures—broken stone, hexagonal blocks, cobblestones— forming a complicated pattern with which the eye could play for hours at a time if it were tired of looking across the Bay. The embankment between the circle and the ramp is planted with rhododendrons, and the concentrated effect of the complex forms and textures, the dynamic rhythm of the space itself, are almost too good to be true. Here abstract geometry, landscape gardening, and architecture, along with the tactile value of sculpture and painting, unite in a deeply satisfying composition. Perhaps it is just as well to leave my account of these great reconstructions at this point. When one has reached a moment of perfection, there is no use pushing any further. But I might add that if there is enough creativeness among New York's park planners to achieve this promenade, with its stirring climax, there should also be enough among our city planners generally to produce a better layout for a complex of public buildings, offices, apartment houses, and public gardens than the present designs for the new civic center have so far indicated.

1953

NINETEEN

House of Glass

For a long time after Lever House opened its doors, throngs of people, waiting patiently in great queues in the lobby, demanded admission so insistently that the elevator system, designed to handle only Lever Brothers' office staff of twelve hundred employees and a normal complement of visitors, was severely overtaxed. People acted as if this was the eighth wonder of the world, this house of glass approached through an open forecourt that is paneled with glistening marble, punctuated by columns encased in stainless steel, and embellished by a vast bed of flowers and—last touch of elegance against the greenish-blue windows and the bluish-green spandrels of the glassy building that rises above it—a weeping-willow tree.

In many ways, this popular curiosity, which in a sense is also popular judgment, is justified. Lever House is a building of outstanding qualities, mechanical, aesthetic, human, and it breaks with traditional office buildings in two remarkable respects—it has been designed not for maximum rentability but for maximum efficiency in the dispatch of business, and it has used to the full all the means now available for making a building comfortable, gracious, and handsome. This whole structure is chastely free of advertisement; the minuscule glass cases showing life-size packages of Lever products in the glass-enclosed reception chamber on the ground floor would hardly be noticed in the lobby of a good hotel. But the building itself is a showcase and an advertisement; in its very avoidance of vulgar forms of publicity, it has become one of the most valuable pieces of advertising a big commercial enterprise could conceive. For years, businessmen vied with each other in the attempt to put up the tallest building in the city; thus the Metropolitan Life capped the Singer and the Empire State capped the Chrysler in the effort to make the sky the limit. In keeping with this now deplorably old-fashioned spirit, there have lately been rumors of a hundred-story skyscraper. Possibly Lever House has pointed the way for a new kind of competition—a competition to provide open spaces and a return to the human scale. At all events, it is definitely not an example of the "swaggering in specious dimensions" that Oswald Spengler called a sign of a decadent civilization.

To understand what the architects of Lever House —Skidmore, Owings & Merrill, whose Gordon Bunshaft was chief designer—have achieved, one must go back to some of the buildings put up on midtown Madison Avenue in the early 'twenties. They are only twelve stories high, without setbacks, and they cover the entire site, providing not so much as an air shaft in the center. But though they have resulted in a heavier density of population than a wise zoning law would permit, they are immensely superior to the extravagant thirty- and forty-story buildings that followed them. So valuable have these older ones proved that one of them, 383 and 385 Madison Avenue, has now been completely renovated and given new elevators, an airconditioning system, and numerous other embellishments at a cost as great as that of the building itself. Lever House returns to the more modest density achieved in this twelve-story structure. By not quite doubling that number of floors in the main part of their building, however, the architects of Lever House

have been able to house those twelve hundred employees comfortably while providing an unusual
amount of open space that is secure against encroachment. For the main structure, though it runs the cross-
town length of the site and abuts the structure next
door on the west, is set back a hundred feet from the
south building line and forty from the north and has
the generous width of Park Avenue to the east. The
result of this self-discipline is that this shaft, or "slab,"
which is less than sixty feet wide, is open to the light
on three sides, and few desks are more than twenty-
five feet from the continuous windows. Even the least-
favored worker on the premises may enjoy the psychological lift of raising her eyes to the clouds or the
skyscape of not too near-at-hand adjoining buildings.
I know no other private or public edifice in the city
that provides space of such quality for every worker.

The layout of this building is itself transparent. The
tall, narrow, oblong slab, which houses the firm's
offices, is set off-center on a roughly square pedestal,
only two stories high, that covers the whole plot, the
western block front along Park Avenue between Fifty-
third and Fifty-fourth. This irregularly shaped site
runs a hundred and fifty-five feet west on Fifty-third
and a hundred and ninety feet west on Fifty-fourth.
The pedestal is a hollow one, for there is a court open
to the sky in the middle of it, just to the south of the
slab. To the north of the court, on the ground floor,
is the glass-walled main lobby, and to the west of the
court are an auditorium and a kitchen laboratory. The
court, and the lobby, can be reached from almost any
direction, for the ground floor is completely open on
three sides—north, south, and east—to the streets;
there is no vestige of wall, or even of shop-front window, to shut out the passer-by. The second floor contains, among other things, an employees' lounge, handsomely done in dark green and mustard yellow, and a

spacious room that houses the stenographers' pool. The third story, the beginning of the slab, contains a kitchen and cafeteria, which can feed all hands in two and a half hours; this dining room, with its reddish-brown drapes and modern furniture, is able to hold its own in elegance with any restaurant on Park Avenue, and it has something that no restaurant in the city has offered since the old beer gardens disappeared—a thickly planted open-air roof garden that flanks it (and, of course, the slab) on both north and south. If it weren't for its almost hepatic sound, the word "Leverish" might well take the place of "ritzy" as a synonym for the last word in luxury. This floor of the slab is indented a whole bay along the Park Avenue side, so the rest of the slab seems to hover over the base of the structure. The indentation permits the bed of plants that borders the roof garden to be carried without interruption along this entire frontage of the building. Unfortunately, the bay is not deep enough to permit people as well as plants to make this journey from south to north. Thus no one can take a full turn on the roof-garden deck, and the architects' sacrifice of free promenade space to the unbroken bed of greenery must be set down as a piece of empty formalism—all the worse aesthetically because the movement of people across the front of the building would have given an extra touch of life to a somewhat glacial, if not oversimplified, composition. This seems to me a blemish, but it is not beyond remedy.

The office building proper ends with the executives' offices, on the twenty-first floor. Above them are three floors, outwardly punctuated by the horizontal louvers of the air intakes, behind which are the elevator machinery and a cooling tank. All this is surrounded by a shell strong enough to support the elaborate machine that moves around the perimeter of the roof to raise and lower the window cleaners' platform.

This piece of apparatus was necessitated by the fact that the entire slab, windows and spandrels alike, is— except, as has already been pointed out, on the west side—sheathed in glass, and the windows are all sealed. The windows are four and a half feet wide, and even the smallest private office has two of them. For a company whose main products are soap and detergents, that little handicap of the sealed windows is a heaven-sent opportunity, for what could better dramatize its business than a squad of cleaners operating in their chariot, like the *deus ex machina* of Greek tragedy, and capturing the eye of the passer-by as they perform their daily duties? This perfect bit of symbolism alone almost justifies the all-glass façade.

The slab is the traditional steel-framed skyscraper, with one or two special features. The outer columns are set back a little from the outer walls, so the windows are a continuous glassy envelope, and the mechanical core of the building—the passenger elevators, the conveyor that delivers outgoing mail to the postal department and incoming mail to the proper floors, the coat racks for the office force, the fire stairs—is concentrated in the west end of the slab. If necessary, therefore, a wing could be built south from this end, parallel to Park Avenue, without taking away any daylight from the existing working quarters. The only opaque feature in this house of glass is that demanded by prudence and the fire ordinances of New York— the fire stairs, which are enclosed in a shaft of light-gray brick at the west side of the site and connected with the slab by open passages at each floor level. At the base of the fire tower is the entrance to the fifty-five-car underground garage for the staff.

Aesthetically, the exterior of this building has a sober elegance; the stainless-steel window frames and spandrel frames are repeated without variation over the whole façade. The darker bands of the spandrels give horizontal emphasis, while the gleam of the vertical metal framing, sometimes reinforced by the columns behind, provides a delicate counterpoise. The effect is of alternating bands of dark-green and light-green glass, and, as is true of all glass buildings, this surface looks far darker than it would if an opaque covering, such as white brick, had been used. Paradoxically, a whole city of such buildings, so open to light, would be somber, since a transparent glass wall is mostly light-absorbing, not light-reflecting. When the framing of Lever House was put up, it was protected by a coating of brilliant chrome-yellow paint, and though the cost of maintaining this brilliance might have been prohibitive, that chrome yellow, playing against the green, would have given the building a gaiety it lacks. Standing by itself, reflecting the nearby buildings in its mirror surface, Lever House presents a startling contrast to the old-fashioned buildings of Park Avenue. But if its planning innovations prove sound, it may become just one unit in a repeating pattern of buildings and open spaces.

The uniformity and the severity of the exterior glass-and-metal envelope do not characterize the interior of the building, for in its decoration this severity has been richly humanized. This décor was designed and executed by Raymond Loewy Associates. Just as a sensible farmer designs his cow stalls around his cow, the fundamental unit around which Lever House's hundred and thirty thousand square feet of floor space was designed was the desk. The desks in the working quarters are of adjustable height and have rounded corners, to reduce the number of nylon snags. To offset the bluish light from the exterior, a grayish beige was chosen as the basic color for desks and floors. (Even the elevator boys are dressed in

dark beige.) But against that background a great variety of colors has been introduced. Each floor has its
own color scheme, from brisk yellows and delicate
blues to a combination—on the floor devoted to the
firm's cosmetics—of boudoir pink and eyeshadow
lavender. I don't know any other building in the city
in which so much color has been used with such
skill and charm over such a large area. Both our
school architects and our equally timid hospital architects have something to learn from this.

There is only one dismal flaw in the excellence of
the interior decoration; this occurs on the topmost
floor, sacred to the chief executives. Here nothing
has been spared to achieve an air of expensiveness,
and as a result nothing more stuffy and depressing
could be imagined. Instead of the clean, shapely
clocks that tell the time on the lower floors, there is
a fussy, ornamented clock, set in a frame of golden
rays; instead of bright-colored hangings and cover-
ings, a drab plushiness, doubtless intended to symbolize solidity, power, and wealth, has the effect of
expressing timidity and the spirit of retreat—in contrast to the forthright confidence and gaiety of the
rest of the building. Why this descent from the era
of stainless steel and glass to the nether regions of
the Brown Decades? Is this a last desperate gesture
toward the good old days, when income and corporation taxes and unemployment insurance and welfare
plans did not exist? The clean logic of the whole
building is denied by this executives' floor. The way
to symbolize leadership and responsibility is not to
give executives a duller kind of decoration than their
subordinates but to give them precisely the same
kind, if on a more generous scale of space.

Because Lever House has many points in common
with the United Nations Secretariat, it is inevitable
that the buildings should be compared. On almost

every point, it seems to me, Lever House is superior. To begin with, it is correctly oriented, with its wide façades facing north and south, and though this means that no direct sunlight ever enters the northern windows, it also means that there is no need to cut light and view on that side by drawing Venetian blinds. Since there are three air-conditioning systems —for the north side, the south side, and the middle —in the winter, warm air can be introduced on the cool side of the building while cooler air is circulated on the sunny side. The United Nations cafeteria for employees is good, but the one in Lever House ranks with the quarters provided not for the U.N. staff but for the executives and delegates. And there is no open space around the Secretariat that compares in charm and comfort with Lever House's courtyard and roof garden, enclosed as these are on two sides.

Few of the features that make Lever House superior are the result of its having a more generous budget to draw on. Though they are superficially similar, one may say of these buildings that the United Nations is the last of the old-fashioned skyscrapers, in which importance was symbolized by height, while Lever House is the first of the new office buildings, in which the human needs and purposes modify cold calculations of profit and nullify any urge to tower above rival buildings. In Lever House, quality of space takes precedence over mere quantity.

The building that Lever House really invites comparison with is quite a different structure, though equally bold and even more striking architecturally in its own day—Frank Lloyd Wright's now demolished Larkin Building, in Buffalo, the paragon of office buildings at the time, though set in the midst of an industrial slum it never succeeded in dominating or even modifying. It, too, was a by-product of the soap industry. In that building, as in this one, every possible innovation was made—new desks, new chairs, new office equipment of every kind, all of it specially designed. The Larkin Building was a shallow structure, built about a great skylighted interior court, with natural light coming down through the roof. Wright's creation was a masterpiece of beautiful masonry—more monumental, in fact, than most public buildings, whether churches or city halls, that have sought to be. Lever House lacks the massive sculptural qualities of Wright's inspired masonry; it is, rather, in its proud transparency, "a construction in space." It says all that can be said, delicately, accurately, elegantly, with surfaces of glass, with ribs of steel, with an occasional contrast in slabs of marble or in beds of growing plants, but its special virtues are most visible not in the envelope but in the interior that this envelope brings into existence, in which light and space and color constitute both form and decoration. In terms of what it set out to do, this building— excluding the deplorable executives' floor and the wall encrusted with golden mosaic that faces one in approaching the elevators on the ground floor—is an impeccable achievement. Lever Brothers and Skidmore, Owings & Merrill, and above all Gordon Bunshaft, are entitled to a civic vote of thanks for taking this important step toward sane planning and building. Lever House is not, of course, the first all-glass building; the famous Crystal Palace, and the more recent Daily Express Building, on Fleet Street, in London, antedate it. But it is the first office building in which modern materials, modern construction, modern functions have been combined with a modern plan. In a sense, it picks up the thread where the architects of the Monadnock Building in Chicago, the last of the all-masonry skyscrapers, dropped it two generations ago.

On the surface, this seems about the best that current architecture can provide when limitations of cost do not, in any substantial way, enter into the picture. It will be a little while before one can make a final appraisal of this building; that will depend partly upon how comfortable the quarters have been in the summer and how expensive it has been to keep them comfortable, likewise on how satisfactory this building will be in very cold weather. It is a show place and an advertisement, and costs that can here be written off to publicity might prove too high for more workaday business quarters. Though the uniform façade of Lever House is aesthetically consistent, a different system of fenestration on the south side, with or without sun screens, might not merely produce better summer temperatures within but might also reduce the need for shutting off the view with Venetian blinds, a necessity that makes nonsense of the windows. It may be, too, that a more flexible system of ventilation, which depended more frequently on untreated air and would use air-conditioning only to counteract extreme temperatures, would prove more satisfactory as well as cheaper. And in that event Lever House's closed-in glass face, along with its amusing window-cleaning apparatus, could be discarded in newer designs. Surely no building so open to the direct rays of the sun—particularly the valuable ultraviolet rays of morning—should nullify that advantage by "windows" that do not let these rays in. But Lever House, by reason of the internal consistency in its design, is at the very least a highly useful experiment. Fragile, exquisite, undaunted by the threat of being melted into a puddle by an atomic bomb, this building is a laughing refutation of "imperialist warmongering," and so it becomes an implicit symbol of hope for a peaceful world. In the

kind of quarters it provides for its staff, Lever House
even anticipates the "Century of the Common Man."
I don't know whether that is what the corporation
had in mind when it built this structure, but that, it
seems to me, is what Lever House itself says.

1952

TWENTY

Crystal Lantern

The Manufacturers Trust Company's new Fifth Avenue office, at the southwest corner of Forty-third Street, revives the dream of building a whole city of glass that haunted the Victorian imagination. Most of the big cities of Europe perpetuated that dream to the extent of building at least one glass-covered shopping arcade; George Pullman, in projecting his ill-fated town of Pullman, Illinois, built a great glass-covered market and community hall; even Ebenezer Howard, most practical of idealists, conceived an enormous under-glass shopping avenue as one of the spectacular features of his proposed Garden City near London; and the dream of sparkling crystalline cities, all glass and metal, figured in more than one of H. G. Wells's many utopias. But only in our time has this dream begun to come nearer to general realization, and among the architects who have done most to put the idea to practical use—Philip Johnson in his own house, in New Canaan; Mies van der Rohe in the Lake Shore Drive Apartments, in Chicago; Frank Lloyd Wright in the S. C. Johnson & Son's Laboratory, in Racine, Wisconsin—none has done more (in New York, anyway) than the firm of Skidmore, Owings & Merrill. In the Manufacturers Trust, they have followed up their Lever House with a quite different mode of design, and in the course of executing it they have pointed up many of the possibilities of glass, both structurally and aesthetically.

While Lever House is, at most times of the day and the night, a dark-green, almost opaque building, in which the glass sheath seems stretched taut, like a film, over the frame it conceals, this new bank is a paradoxical combination of transparence and solidity —crystalline, yes, but not in the slightest frail or film-like, for, if anything, it is both rugged and monumental. Lacking any vestige of classic Greek or Roman form, or anything to remind one of the conventional banking temple except its low, four-square form, it nevertheless expresses the classic qualities of dignity, serenity, and order. To the observer approaching it along Fifth Avenue, this structure, with its heavy vertical aluminum ribs and its massive glass wall panels, conveys the feeling that it is there to stay. The over-all treatment of the façade actually says aloud what the engineers' calculations had indicated —that these glass walls could withstand a wind of up to a hundred miles an hour, a velocity that, because of its sheltered position, the building would never be subjected to.

The notion of designing a bank with an all-glass façade was first broached, I believe, in these columns in the nineteen-thirties, the theory being that exposure to public view was a much greater protection against assault and robbery than any number of stately Corinthian columns and bronze grille cages. As a matter of fact, the classic column, once the pat symbol of conservatism and financial solvency, did not, as a symbol, survive the awful financial revelations of the great depression, yet it was a long time after that before banks began to free themselves of this antique image. The move toward modern design in banks first manifested itself only in the interiors; the cold, chilly, railroad-station atmosphere was gradually warmed up with "homey" touches of the sticky kind associated with Georgian revivalism, and

later with equally dubious attempts at mural painting
in the genteel tradition of the eighteen-nineties. During the past ten years, though, many banks—there is
a whole nest of them around Rockefeller Center—
have adopted modern decoration of the sort one
might expect to find in a fancy club. But nowhere
have both interior and exterior been conceived more
effectively as a whole, or treated in a more forthright
manner, at once businesslike and elegant, than in this
new structure. The great merit of the Manufacturers
Trust's new quarters is that, being all of one piece,
every part tells the same story, and to perfection.
This is true of the little things as well as of the big
ones; thus, the clock on the second floor, ignoring
the foolish modern convention that requires blank
dials and small, stubby hands that don't indicate more
than a vague approximation of the time of day, actually tells the hour with pointed precision. Admittedly,
part of the building's success is due to the very special advantages of its site, which I will more fully
touch on later, but the fact that the architects made
use of these advantages is no little part of their
achievement.

Viewed from outside, this building is essentially a
glass lantern, and, like a lantern, it is even more striking by dark than by daylight. The main shell contains
four floors, counting the street-level one, and it is
topped by a fifth floor, a penthouse that is set back
to provide a terrace, which, like the Fifth Avenue
frontage, is planted with trees, thus suggesting the
natural landscape so many of us leave behind when
we must do business in the Plutonian underworld
of megalopolis. This landscaping makes the building
an urban equivalent of Poe's "Domain of Arnheim."
Indeed, the bank is as prodigal in the luxuriance of
its vegetation as in the luxury of its working equip-

ment, for the extensive greenery, planned and installed by Clarke & Rapuano, landscape architects, requires the services of a gardener two full days a week. The glass walls of this building—there are no windows—are set between tall, vertical ribs of aluminum that rise from a scuffle base of black granite. These walls are interrupted only by three emphatic spandrels of dark-gray glass that girdle the upper floors, while the second story, a mezzanine, is fringed by a thinner band of leafy plants, which offer a contrasting touch of green. The crystalline quality of the structure is further brought out by sheathing the ceilings in opaque plastic panels, ribbed like a washboard, behind which cold-cathode lamps evenly diffuse a pale-yellow light throughout every floor. The only screen of any kind against sunlight is provided by spun-glass curtains of neutral color and texture, which are probably more useful for insulation than for visual protection. The glass panels are really curtains themselves, for although they are firmly fixed in place, their crushing weight is actually suspended from the roof, which, like the floors, is cantilevered out from the columns that support the entire building. The suspending of these walls eliminates the need for heavy, clumsy supporting beams, and thus adds to the feeling of lightness in the structure. This is but one of the many costly technical feats that make this whole audacious enclosure of space seem so simple and so effortless. Another trick is using only eight of these columns, all of them freestanding and all of them set well back from the outer walls of glass, to support the building. This device, which opens up the interior space by eliminating the once mandatory row upon row of supporting columns, is one of the positive contributions of modern engineering to the luminous, lantern-like effect of this building.

To the passing spectator, there are two features

that become striking only when he nears the building. One is the great horizontal banks of foliage in the huge boxes that serve as pedestals for the pair of rapidly moving escalators that rise from the main floor to the mezzanine on the Fifth Avenue side. The other is the vast round entrance, also on the Avenue side, to the vaults of the bank, with its massive door of shiny metal—the most impressive possible symbol of security. This thirty-ton door holds the eyes of a constant crowd on the sidewalk, as once upon a time the white-capped chef making hot cakes held people in front of Childs. (Customers with business in the safe-deposit vaults enter them through a series of less ostentatious but well-guarded portals from a small side door at the western end of the Forty-third Street façade.) By raising the most dramatic physical object in a bank from the cellar to the ground floor, the architects have made the most of a natural advertisement. This is what one might call inherent symbolism; it contrasts sharply with the more traditional kind, as old as Assyria—a symbolism that might be represented by two ferocious granite lions. This use of the bank's vaults as an expressive and visible feature was truly an inspiration.

The main entrance to the bank is on the Forty-third Street side, near the corner, and it is clearly indicated by a vertical slab of black granite in the middle of it and a horizontal siding of the same stone, with the bank's name, above it. There are two even more modest legends, in stainless-steel letters, at eye level on the glass of the Fifth Avenue façade. Remembering how one of Louis Sullivan's little banks in Iowa was bedeviled by no less than seven signs, all frantically proclaiming the name of the bank, one is grateful to the Manufacturers Trust for believing that this is enough; after all, the building has already become self-identifying.

At last we are ready to enter the building. On the ground floor, the main aesthetic ingredients of the whole composition are on display—the ceiling of corrugated plastic exuding that even yellow light; the pale terrazzo floor; the open tellers' counters, minus the old grilles and cages, forming an ell against the back walls; i.e., the south and west ones. The first of these walls, which also serves as an outer wall of the bank vault, is of black granite up to the mezzanine, and then white, light-green, and blue plaster on successive floors; the other is of sky-blue plaster up as far as the mezzanine, and then gray marble. (The walls of the fifth, or penthouse, floor are all of glass.) Since the escalators go no higher than the mezzanine, there is a bank of elevators, with handsome red-lacquered doors, in this wall. The bases of the tellers' counters are of ebony wood set between vertical ribs of stainless steel. The tops of the counters are done in a beautiful, creamy Italian marble, with a light-tan figure weaving through it. The ebony of the counters is repeated in the long, bowed table in the directors' room on the top floor, and the marble of the counters is used again for the floor of the president's room, and as a wall panel on the fifth floor. This flow of materials and colors from one floor to another, now for one purpose, now for another, contributes powerfully to the unity of the building, while the basic palette used by the design consultant, Eleanor Le Maire, is sufficiently comprehensive—white, light yellow, tan, brown, blue, and black are her principal colors—to be capable of endless variations and combinations. Radically contrasting colors are introduced only in the powder room, on the fifth floor—where all sorts of lipstick pinks and reds dominate—and in the employee's lounge, in the basement, where, doubtless in an attempt to be gay and to put banking associations aside, the color scheme is somewhat distracting, if not incongruous. Apart from this lapse, and the fact that the sky blue of various wall spaces is not merely cold but is at odds with the indigo blue in some of the textiles and leathers, Miss Le Maire's decoration seems in admirable accord with the spirit of the whole structure, including its directness—one might also say its masculinity.

Aesthetically, the topmost floor and the main banking floor are almost without reproach. The first of these is the final word in quiet luxury, though it utilizes no materials or motifs that are not already at work on the other floors. The most notable part of its decoration is the perilously artful arrangement of potted tropical plants (such as philodendrons) whose waxen perfection rouses a suspicion of complete artificiality, and the series of modern prints, paintings, and sculptures that bring into this highly rationalized interior some of the more subjective emotional elements that are usually absent from the surface operations of a banker's mind, and certainly absent from the kind of art that banks have in general patronized. But the second floor, dedicated to the senior officers of this branch of the bank, and to the Trust and Foreign Departments, is the crown of the architects' and decorator's aesthetic achievement. The noble height of this story, the tallest of the five, emphasizes its almost unbroken space, and the cold pallor of the white walls and the white marble-sheathed columns is softened by a splatter of vermilion chairs, in a reception area near the escalators, and the brilliance of a huge golden screen that is placed near the west end of this floor to separate the counter in front of it from the elevators behind. This screen is a feature that I gravely doubted when I saw the preliminary illustrations, but the truth is that it lifts the whole composition to a higher plane. It is a mass of loosely assembled and variously twisted oblong steel plaques, plated with bronze and gold and welded together, and it has just enough minor variations in texture and color, plus an occasional break into abstract shapes, to make it stand out from the glossy mechanical refinement of the rest of the structure. The screen is the work of Harry Bertoia, the Pennsylvania sculptor. Though it is purely abstract, making no effort at symbolic significance, it humanizes these quarters even more effectively than the living plants, mainly because it suggests something frail, incomplete, yet unexpected and defiant of rational statement, and thus lovable, a note that is not audible in most of the representative architectural expressions of our time. On this ground, one might also defend the tangled hair-net of wire that floats in the air above the escalators as they end, on the second floor, and that was meant to cast a complicated shadow on the south wall, but, frankly, this creation does not seem worthy of the sculptor and craftsmen who executed the screen. And the shadow has evaporated in the diffused glare of the overhead lights, since the spotlights that were to produce it were too strong for the tellers' eyes. (This problem, however, is being given further study.)

All in all, then, the Manufacturers Trust Building is perhaps as complete a fusion of rational thinking and humane imagination as we are capable of producing today. If you reject it, you reject many of the notable excellences of our age. As a symbol of the modern world, this structure is almost an ideal expression The interpenetration of inner space and outer space, the fact that the principal functions of the building are as visible from the outside as those of a supermarket, that the same freedom of space and light has been provided in every part of the structure, thus giving the executive officers, the staff, the clients the same architectural background—this surely reflects the economic, the social, and the aesthetic principles of the modern business world at their best. This architecture is a formal expression of the culture that has explored the innermost recesses of the atom, that knows that visible boundaries and solid objects are only figments of the intellect, that looks with the aid of X-rays and radioactive isotopes at the innerness of any sort of object, from armor plate to human bodies. That architects and sculptors should seek to express this world of transparent forms and dynamic processes, that they should actually revel in this exposure, is natural, since it is the eternal business of the artist to express in new forms what would otherwise remain at the commonplace levels of daily perception.

But there is a catch to this method of symbolic interpretation that some of our best architects have too often overlooked. Achieving effective symbolic expression frequently demands a surrender of practical convenience to expressiveness, a sacrifice acceptable only for religious ends. An all-glass building, fully exposed to sunlight, is a hotbox in our climate, and then the penalty one usually must pay for symbolism is to cower a large part of the time behind a Venetian blind, which robs the form of its aesthetic significance. Happily for the architects of this bank, its location and its purpose greatly reduced the practical objections to an all-glass structure. Because it is surrounded by taller buildings, there is no need for screens or blinds or green heat-ray-resistant glass. In addition, the architects, instead of treating all the functions of the bank as public ones, because of a theoretic commitment to the open plan, have provided a series of snug rooms, with doors, on the third floor to insure privacy where privacy is desirable. In other words, the functional requirements of this build-

ing coincide remarkably well with the symbolic or
expressive functions, and the result is a bold, straightforward design, consistent in every detail without
being arbitrary or formalistic. But let novices take
warning! An attempt to imitate this design under different physical conditions might easily result in lamentable sacrifices of pleasure and comfort.

And this brings me to the last factor that insured
the excellence of this structure: no expense was
spared in its conception and execution. By making
this building just large enough for its own uses, instead of pushing it up to a height of ten or fifteen
stories to provide pigeonholes for other tenants, the
bank not only refrained from adding to the congested
population of Fifth Avenue but enabled the architects
to conceive a completely articulated and unified structure. Some of the best features of its design would
have been forfeited if it had been necessary to build
the usual steel skeleton with the usual number of
columns. Again, the system of ceiling and cove lighting chosen was an expensive installation, but it may
be that because of its unity, efficiency, and pleasant-
ness it will turn out much cheaper in the long run.
In other words, the richness and perfection of the interior are due largely to the fact that there was no
penny-pinching about first costs. Perhaps the bank's
president, Mr. H. C. Flanigan, took to heart the lesson
of this bank's only architectural rival, the handsome
P.S.F.S. Building, in Philadelphia, done by Howe &
Lescaze in 1932. For there the costly materials and
meticulous craftsmanship have, in the course of a
quarter century, paid off in utility, low upkeep costs,
and undatable beauty. What matters, really, is not first
costs but final costs. Possibly the cheapest quarters in
the world, as regards *total* cost, and the most profitable
in the long run, are the group of buildings—originally
residences—that frame the Place Vendôme in Paris.

They are still spacious and sound enough to be serviceable (and bring high rents as offices) after two and a half centuries. By thinking in such long-range terms, more New York enterprises might produce buildings and even whole districts of comparable excellence.

1954

TWENTY-ONE

Windows and Gardens

The finest piece of new architecture New York has
seen since last year's Frank Lloyd Wright retrospec-
tive show belongs, unfortunately, like the admirable
prairie dwelling house that show contained, to an
ephemeral species, the museum exhibit—here today
and gone tomorrow, sometimes before one has even
heard of its existence. Happily, thousands of New
Yorkers responded to the brief opportunity to examine
the Wright showing, and these (or other) thousands
have lately been visiting the Museum of Modern Art
to examine its Japanese house, done in the traditional
manner, which the Japanese themselves are rapidly
losing their grip on. One can't help wishing that our
museums wouldn't take Ben Jonson's reflection that
"in short measures life may perfect be" as a command. I can think of no more useful demonstration
of the nature of architecture and the origins of the
modern movement than these two houses would have
offered if the spectator could have moved thoughtfully
and at leisure back and forth between the Modern
Museum's handsome production, which includes a
Japanese garden and is done with the utmost care
and refinement, and Wright's house at the Solomon R.
Guggenheim Museum, even though it was on view at
a place and a season that did not favor the garden
accompaniment.

Lest any deluded soul should be tempted, out

of passionate attraction to this beautiful building, to order one for himself on Long Island, as a successor to the fading ranchhouse style, certain things should be pointed out. This is not the first Japanese house in the traditional manner to be put up in this country. Even before the Chicago World's Fair of 1893, which contained a Japanese exhibit that may well have influenced the young Frank Lloyd Wright, a returning missionary had built one in Rhode Island. This was part of the first wave of Japanese culture, which John La Farge and Ernest Fenollosa, the nineteenth-century connoisseur, brought back from Japan at the moment the Japanese were showing a hideous strength in their attempt physically to Westernize themselves. When the Japanese forms reached these shores, the interior of the American home, everywhere except (perhaps) in the more puritanic purlieus of New England, had become a combination of an Old Curiosity Shop and an Atlantic City auction room, writhing with all manner of corrupt bric-a-brac and decoration, one ugly object concealing another in the name of art or "hominess." Clutter was mistaken for culture.

More than one kind of tool was needed to remove this aesthetic debris, which had accumulated so rapidly with our increasing wealth and our increasing facilities for mechanical reproduction—to say nothing of our increasing exuberance of bad taste—after the Civil War. New standards of hygiene influenced this transformation; for example, Florence Nightingale's white hospital rooms and her own white sitting room, into which light and sunshine entered through the windows with a minimum of interference, had a definite effect. But the lesson of the Japanese house, scattered widespread by the Japanese print, was probably as powerful as any other force, for it associated purification and cleanliness with beauty. Beauty, the

Japanese proved, was not something that one added
to an already finished structure; it was like the work
of the carver, who took a rough, unformed block of
wood or stone and by a steady elimination of the su-
perfluous produced a form that had outline and character. All this is excellently demonstrated by Junzo
Yoshimura's house at the Modern Museum. Its quality lies in the kind of space it produces, and in the
subtle changes in the color and texture of the surfaces as one's eye passes from bamboo to pine, from
pine to cryptomeria or cypress, and from the wood
surfaces to the paper of the screens or the straw mats
on the floors. This type of Japanese house is a rural
product; its origin is the primitive Japanese farm-
house, which was in effect merely a gabled roof
rooted in the ground, and even in its utmost refinement of construction it remains essentially a wooden
building in every detail, its curved roof covered with
layers of cypress bark, and its interior columns, though
the bark has been peeled from them, bearing the raw
imprint of nature in every curve and knot. By the
thirteenth century, Japanese paintings show us, this
house had attained its basic modern form; indeed, in
both their rambling plan and their low-pitched roofs,
the early houses are closer to what delights us today
than the steep-pitched roofs and overelaborated interiors of the houses of the middle period. This
Japanese house, with wide overhangs and generous
verandas to encourage open-air living, functions admirably in its humid native summer, but there is
ample testimony that it is too frail to give protection
against sharp winter weather or prevent the passage
of noise from room to room. Partly to make up for
this last, though, the plan of many Japanese houses
—and this is true of the version at the Modern Museum—separates its various functions into wings, with
passageways between. Like most Oriental dwellings,

the Museum one is screened from its neighbors and
the street by a wall. The only point on which I question it is the use of plastered masonry, topped by
curved gray tiles, instead of wood to form the wall
that serves as the background of the garden, but I
have no doubt that there are many precedents for
this. One would be the masonry walls of the old Kyoto
palace. All in all, however, this is a consummate
work.

The greatest lesson the Museum house teaches is
how much beauty can be achieved merely by quiet
repose, by selection and elimination, by stripping
every human requirement down to its essentials. To
sit at ease on the floor is the act of an athlete, and
that kind of athletic facility is demanded everywhere
in the use and enjoyment of the Japanese house. The
simplification that the Japanese architect seeks is not
a mechanical process; it owes a heavy debt to the
seemingly mythical Taoist sage Lao-tse, who observed that it is the hollow that makes the bowl. If
simplification were not at the service of function *and*
form *and* idea, it would have no more aesthetic significance than the Yankee process of whittling, whose
final result is only shavings and splinters. Many architects think they have achieved a modern form merely
because they use plastics and steel and don't waste
any of their clients' money on ornament. They mistake lack of imagination for "the contemporary
touch," and they take refuge in the fashionable slogan "less is more" without realizing that their particular less is less indeed. But the fact is that the more
one eliminates, the more important it is to refine every
detail and to measure with the eye every proportion
as meticulously as Mies van der Rohe does. The
Japanese have had long experience with these refine-
ments, and they have imposed a rigorous order upon
their floor plan by making the length and breadth of

their rooms a multiple of the length and breadth of
their standardized floor mats. The dimensions of
Japanese rooms are given in mats; for example, the
main room of the Museum house is a fifteen-mat one.
(Our current slang for this is "modular construction.") As a result of this formalism, this rigor, every
object that departs from it—even a vase, a flower, an
orange silk cushion—has an intensity of visual impact it would never have in a more sloppily conceived interior. Because of this simplicity, the little
changes of texture and color and the play of lights
and shadows among the beams, the contrast between
low doorways and high ceilings or between outdoor
brightness and indoor sobriety, have an enlivening
effect. Above all, the Japanese house demonstrates
that purity and simplicity are the aesthetic flowers of
a life that is conceived from first to last in these elemental terms. Just as you must take off your outdoor
shoes before you enter such a house, you must eliminate many other incongruous habits, including a
fondness for gadgets.

So different from the culture of our day was the
whole concept of the culture which produced this
house that even the contemporary Japanese, finding
themselves not fully worthy of these aesthetic refine-
ments, tend to adopt more commonplace modern
forms. Nevertheless, the marvel is that so much of
the Japanese house has already been absorbed by
American architects, beginning with the Greene brothers, Henry and Charles, half a century ago, and
Frank Lloyd Wright, whose passion for Japanese art
has been an equipoise against his admiration for bold
and even brutal masonry, like that of the Mayan civilization. The Japanese love of nature, reinforcing
our own native romanticism, helped win us away from
the paint-and-plaster finishes that used to be standard, and has taught us to take pleasure in natural

wood and stone. The Japanese demonstrated, too, the convenience of the one-story house, or bungalow. Though the open plan had begun to characterize American domestic architecture in the 'eighties, long before Japanese plans could have been widely copied here, the Japanese precedent not only strengthened this tendency but helped to break down the too static division between indoors and outdoors; the sliding screen door of the Museum house opens up a whole vista of the garden from within. In the example in the Modern Museum, the little interior "patio" garden shows how much living beauty can be encompassed in a space less than half the size of the conventional New York twenty-by-sixty-foot back yard.

Yet this traditional Japanese house has points the American architect has been painfully slow to recognize. Despite his extrovert delight in the open plan and the multi-purpose living room, he has not made full use of the sliding doors, either to furnish privacy, when that is needed, or to give a sense of snugness and enclosure, as an escape from the nondescript emptiness of the big living room when it is not housing a large group of people. Few modern American dwellings are as flexible in their employment of space as the traditional Japanese one. Again, the American architect, in his pleasure over the idea of "bringing the outdoors inside," has created fixed glass walls, which not merely filter out the healthful ultraviolet rays of the sun, along with the perfumes of the garden, but leave the householder no choice between complete visual exposure and complete enclosure behind blinds or curtains. This sort of design lacks the Japanese visual contrast between light and dark, as well as the psychological contrast between inner and outer. Such indifference to visual contrast, such disregard of privacy, indicates a certain coarseness of feeling in the American architect, which is another

way of saying that he is the victim of his own mechanical formulas. Not the least merit of Frank Lloyd
Wright's houses is that, probably by instinct, he has
followed the same processes of thought as the classic
Japanese architect—thus his use of wide overhangs
and his interiors, which, even when they embrace a
garden, have their own innerness.

To be at home in this Japanese house, two things
are requisite. One is a love of the simple life as deep
as Thoreau's at Walden. (With this goes a positive delight in the formal expression of simplicity.) The
other is plenty of cupboard and closet space for goods
that are not necessarily beautiful and have ceased,
for the moment, to be useful. Strangely, this is one
of the last things the American architect has learned
from the Japanese; only during the last decade has
he begun to supply all the storage room, behind sliding doors, that is essential to run a house without creating visual confusion.

Another lesson the Japanese house should under-
score for the American architect and client is the
true and proper use of the window. The modern designer, believing that glass is the one veritably modern
material, has tended to equate the degree of his modernity with the number of square feet of glass he
provides, and he has thus created some of the most
intolerable hothouses and cold frames that have ever
been offered as permanent homes for human beings.
A house that will compensate for the harsh extremes
of the American climate needs not a solid expanse of
glass but a multiple set of screens—a sliding wire
mesh to keep out insects, a sliding glass screen to permit a view but to keep out wind and rain, and a sliding
insulating screen to exclude heat and cold as well as
noise. To carry out these requirements, the American
planner will have, I am afraid, to give up his opaque

passion for the transparent wall and go back to the alternation of solid and void that is characteristic of the Japanese house.

A major delight of the Modern Museum's house is that its boundary wall encloses a romantic garden in the Japanese style, even to the rocks and falling water —a garden to be seen from the porch, as a backdrop of nature, not to be used for stretching one's legs. This garden serves as a Buddhist image of paradise, and it is a fine example of the Japanese faculty for creating a miniature world, almost overpowering in its variety of form, which, because of its miniature scale, seems as rich in contrasts as several rambling acres of natural woodland.

One's happiness at seeing this exemplary Japanese garden preserved is increased by the fact that it has a worthy American companion in the new Abby Aldrich Rockefeller Sculpture Garden, which adjoins it. This is a commendable example of contemporary architecture and garden art, though of a more meager aesthetic province. It serves a manifold purpose—as an outdoor display space for sculpture (always best seen under natural light), as a resting place for those who have acquired "museum fatigue" or the "Surrealist blues," and as a foreground for people eating outdoors under the rows of trees and umbrellas of the new restaurant at the west side of the garden.

The space is rectangular, bounded on the west by the high blank walls of the Museum restaurant and the Whitney Museum, on the south by the windowed walls of the Museum proper, and on Fifty-fourth Street by a high wall with two wide wooden gates. These gates would be justified if only because they form a visual break in the mottled-gray expanse of variegated brick, the same in color if not in bond as the northernmost three panels of the Whitney Museum. Two rectangular watercourses (no falls or water lilies) parallel the street wall, and a central white marble plaza, sunk below the level of the Museum and restaurant terraces, forms a solid oblong, bordered on three sides by a green embankment of myrtle. At either end of this plaza are irregular geometric patches of green. The one to the west is the base for a clump of lovely tall white birches; directly in front of the Museum are two irregularly formed weeping beeches, while on the right is a group of Japan cedars, and there is even, farther to the rear, a tree of heaven.

The architecture of this garden, severe and formal, is an excellent pedestal and background for the sculptured figures, especially those in bronze, and the living part of the garden provides an almost romantic contrast with its variety of exotic species. My only criticism of this diversity of tree forms is that it offers a competition to the sculpture that the narrower range of foliage and outline in the more conventional formal garden would not present. But the total effect of this garden, designed by Philip Johnson and landscaped by James Fanning, is a happy one, and it gives the buildings of the Museum a grace and completeness that they had sorely lacked.

The contrast between the two types of garden, the Japanese and the American, is the contrast between wild nature—captured as in a zoo without visible cages, then concentrated and transformed into art, and further enriched by ideological symbolism—and an outdoor room in which nature is served on a pedestal, as if it were itself a piece of sculpture. The final lesson, taught by both the Modern Museum's and the Japanese garden, is that no building is aesthetically finished until the space about it is not merely open enough to permit one to see the building but is transformed by art to the point where it carrics the order and delight of the interior outward and back in again to the observer's roaming eye.

1954

TWENTY-TWO

Museum or Kaleidoscope?

At 22 West Fifty-fourth Street stands the new Whitney Museum, a four-story building on a plot that was once the western end of the Modern Museum's open-air garden, and for once the whole address has been written out above the entrance in clear white letters —a boon to all the out-of-town visitors who often do not know in exactly what part of the city they are. The Whitney abuts at right angles on the Museum of Modern Art, which is to the southeast, facing on Fifty-third Street, and just across the way from the Modern Museum, on the south side of Fifty-third Street, is the new Donnell branch of the Public Library. Together, these three buildings could have formed a cultural center whose visual unity would have underlined its functional unity if only someone had been foresighted enough, back in the 'thirties, to buy up the litter of dilapidated mansions that adjoined the sites, for with even this little extra space the buildings could have been set apart from their neighbors by a continuous girdle of gardens. Most of our dreams come belatedly, when the price of land is too high. In general, therefore, our public authorities and public-minded citizens, whenever they embark upon a civic project, surround it with open spaces that are frightfully and disproportionately expensive, or else provide none at all. But the sort of thing that Mr. Robert Moses did during the depression, seizing upon

odd bits of land that had come to the city in default
of taxes, or that were no longer needed by it, and
turning them into tiny neighborhood parks or playgrounds, should be going on constantly, whenever
opportunity offers. In time, these little patches could
change the texture of the whole town. One of the lessons that even Cinerama tourists might take home
from Venice is the importance of a sudden opening
of color, sunlight, and greenery in a crowded city. We
may have to wait half a century for a large-scale rebuilding of central Manhattan, but meanwhile every
blighted building site that is translated into public
open space would redeem, by just that many square
feet, the foul mess that our Cyclopean (one-eyed)
financiers and builders are making of the midtown
district.

Fortunately, though, the Whitney Museum is closely
tied to the Museum of Modern Art both structurally
and visually, and while its only private garden space
is the tiny terraces at the rear of the second and the
fourth floors, its eastern façade flanks the Modern
Museum's garden. By this close relationship, both
Museums have architecturally profited. The original
Whitney Museum, on West Eighth Street, was a group
of old houses remodeled into a single building, which
bore many traces of the original layout. The architect of this transformation, Auguste L. Noel, did a
tactful and on the whole ingratiating job, keeping the
new building in harmony with the belated Georgian
of Greenwich Village, and Bruce Buttfield gave it an
unobtrusively modernized, if not modern, interior.
The exterior of the present Whitney was designed by
the same architect, working in consultation with
Philip Johnson, who designed the addition to the
Museum of Modern Art (it is this addition upon
which the Whitney abuts) and its garden; and the interior was done by the same decorator. Even if the

Whitney trustees had not felt a natural obligation to conform, there was good reason for them to reciprocate the Modern Museum's generous donation of a fine site by seeing that their architect designed a building that would harmonize with its neighbor. On the east wall he has done this admirably, using windows framed in black, and exposed vertical steel columns, in the fashion of the Modern Museum's addition, which (by the way) is now almost blotted from view by the Whitney building. By using the same grayish-buff brick that walls the garden he has unified the Whitney with the garden and reinforced Mr. Johnson's own design.

The best approach to the Whitney Museum, accordingly, is not along Fifty-fourth Street but from the Modern Museum's garden side. As many people must by now have discovered, the famous open-air public restaurant of the Modern Museum has been preserved (under glass, as it were) despite the arrival of the Whitney; it now occupies the east side of the Whitney's ground floor, and its great glass windows, ceiling-high, give on an open-air terrace, lined with tables and sun umbrellas during seasonable weather, that overlooks the Modern Museum's garden. This horizontal bank of windows is matched by one of green glass along the top floor of both the Whitney and the Modern Museum's addition. Indeed, were it not for the ticket booth of the older Museum in the passage between the Whitney and the restaurant, one might wonder where one building ends and the other begins. Between these banks rises a serene expanse of that grayish-buff brick, broken only by the vertical lines of steel that divide the wall into bays. (The second and third floors of the Whitney contain galleries; the fourth is devoted to offices.) The style of this extremely simple façade, like that of the Museum garden, reflects the aesthetic chastity of Mies van der Rohe. With the slim birches in the garden to throw their shimmering green leaves against it, the façade is the right background for the garden, and the garden in turn is the right foreground for the plain glass and brick and metal wall, since without the tracery of trees that solid wall might be formidable.

From the outside, the façade is a quiet but indisputable success, marred only slightly by the finishing of the rooftop housing of elevator and ventilator shafts with brick tinted a pinkish orange. On the inside the great success is the top floor, devoted to the administrative offices, which occupy its perimeter on three sides—quarters in which the secretaries' smaller rooms share the light and view enjoyed by their bosses. Except for the ornate details of the Trustees' Room, the decorator and the architect are at their best here, and the view from the garden side is so striking—the right word, in fact, is "beautiful"—that one wonders why the entire wall on this side was not made of glass. In that case, it is true, it might have been hard to get people to look at the pictures. Still, a museum with a shallow interior and with light available on three sides would seem to cry for natural lighting, and one's appraisal of the interior must bear this fact in mind.

I wish I could stop at this point, for my admiration would then be fairly copious. But the only other side of the Whitney that is visible, its north façade, fronting on Fifty-fourth Street and containing the entrance to the building, is not quite so successful as the east wall. If one could approach that entrance head on from a distance, the main features of this façade would register well—the sober brick wall, the black-rimmed upper bank of windows, the highly conventionalized black-and-gold eagle flanking the huge metal legend "Whitney Museum of American Art," to the right of it. Unfortunately one must approach it along Fifty-fourth Street, and neither the eagle nor the lettering, which are flush with the wall of the building, are legible, even from the opposite side of the street, until one is quite near. At that point, one's attention is distracted by an incongruity of detail on the street level: the stark panels of oyster-white mosaic on each side of the entrance, the black metal of the service doorway, and the black frame of metal and highly polished marble that rims the mosaic. Not merely are the white panels themselves too blankly emphatic, despite the texture of the mosaic, but this combination of white and black has unfortunate associations; it gives the building a Funeral Home elegance, if it does not, indeed, unhappily remind one of the black glass used for the façades of cheap shops and provincial movie palaces. That in the Whitney Museum these materials seem rich and costly does not make the result more attractive. The eagle, which Mies van der Rohe would surely never countenance, could perhaps be assimilated by this austere design, but the patches of white mosaic, whose texture and color contrast so sharply with the brick wall above, are flatly indigestible.

By this treatment of the entrance level of the Museum, the architect unfortunately set a precedent for the decorator of the interior, and it is hard to give an account of that interior, and above all of the lobby, without seeming disrespectful. The space that one faces upon entering is divided into two parts— the lobby proper and a sculpture gallery, a few steps down, beyond. On the right of the entrance is the cloakroom, walled with opaque glass that has been textured into little cubes and set in black metal; on the left is the usual counter, where catalogues and books are sold. Beyond the cloakroom, and again on the right, is a bank of elevators, framed in polished marble, and beyond them is an open stairway that turns a corner in its ascent to the second floor. Beyond that is the sculpture gallery, whose walls of light-gray brick form a pleasant background for the sculpture. Here, however, is a place where the use of a stronger, deeper color on one wall might have brought out with greater effect the sculptures that are done in light stone.

The elements of this whole composition are so simple that it is difficult to go wrong with them; in fact, in a black-and-white photograph this entire floor seems quietly effective, especially because, except for the coping and framing of the stair, the forms are clean. But the designer has overcompensated for this restraint by using so many materials and colors and textures that the place looks like an architects' sample room. There is black metal around the entrance doors and the coatroom, a gold-metal frame around the showcases behind the counter, transparent glass in the entrance doors, opaque glass for the cloakroom wall, and the floors and the other walls involve terrazzo, brick, mosaic, plaster, and marble, in as many colors as textures. A bust of Mrs. Whitney is embellished by a setting of five kinds of stone, and two kinds of marble are used to empanel the elevator doors. A rectangle of yellow needlessly draws the eye to the stair well. Despite this profusion of colors, the effect is dull, as in a variegated fabric whose hues have been washed out by too many launderings. Instead of accentuating forms, these colors cancel out and make the room poorer than it would have been if a more limited palette had been used. And despite all this labor to achieve animation and contrast, the one place contrast would have been effective was overlooked; the terrazzo floor unites the lobby and the exhibition gallery, whereas a change from a dark floor to a light floor would have defined the two spaces clearly and emphasized the difference in function.

Though these are individually small faults, they add up to a big one.

Much effort and thought doubtless went into this scheme of decoration. The decorator, one guesses, was probably irritated—who by now is not?—by the Mies van der Rohe slogan "Less Is More," which every junior draftsman has learned to parrot; he may have revolted against the puritanism of that formula and felt challenged to find an equivalent, through the use of materials alone, for the involved ornamental forms that were once the special province of the decorator, at a period when "more" meant the works. Except for a few slight touches, such as the volute termination of the stair rail, he has obeyed the law of modern design to the letter, but he has defied its spirit; and though one may sympathize with his resentment, one cannot praise the result, since these muted colors and spotty shapes are irrelevant and distracting. In an art museum, this offense is a particularly grave one; the best counsel for the architect or decorator is to be as self-effacing as possible, and to let the paintings and the sculptures speak for themselves. Even a ponderous classic mausoleum like the National Gallery in Washington has this virtue in no small degree, for its marmoreal forms are so conventional and so anonymous that they are unnoticed, and create only a happy impression of great light and space. By contrast, the brocade-lined rooms of the Metropolitan Museum's recent renovation are much less pleasing—to anyone who loves pictures more than brocades—than the rooms with plain painted walls. The visual emptiness of a good museum or gallery—Stieglitz's 291 and his An American Place are examples—gave considerable aesthetic support to the modern movement in the first place; pictures, it turned out, looked better against a simple background than they looked amid the intricate decoration of a baroque palace. This offered a special justification for the grand clearance of "furniture" and *"l'art décoratif"* made by Le Corbusier and Gropius in the early 'twenties, and it is not an accident that the painters—from Whistler to Ozenfant—had led the way.

It is a pity I cannot add that the second and third floors, which are dedicated in general to painting, redeem the unfortunate impression made by the entrance hall. But the smoking alcove—a pleasant feature that even old museums are now installing—that greets one at the head of the first flight of stairs has the same colorful but unimpressive fussiness. As for the picture galleries themselves, they present some problems of a general nature that have never been satisfactorily solved. If I raise them here, I cast no special reflection on the architect and the decorator of the Whitney—and the Museum's staff—for their own particular solution, since any answer rests on matters of taste that involve one's whole philosophy of life, and on such points good men may well differ.

The galleries are a series of rooms separated almost entirely by freestanding removable partitions and lighted almost wholly by artificial illumination, for the only windows on these floors are at the south end and the glass wall on the second-floor terrace is heavily curtained. Since this is a small museum, with rather low rooms, the color and the texture of the walls, the ceilings, and the floors have a very direct impact on the beholder, and my judgment on them is, I regret to say, unfavorable. The floor is of a wood composition tinted a drab gray-green, a particularly dismal color under the cold lighting, and divided into small squares by metal strips whose shininess unhappily calls all the more attention to it. As for the fabric that covers the walls, its coarse texture and uneven painting are likewise remorselessly brought

out by the glare of the lighting, and whatever its mechanical merits as a medium for hanging pictures, its
effect, too, is dismal. But these flaws are nothing compared to the effect of the illumination—cold-cathode
and fluorescent lights concealed behind a glass ceiling that, like the floor, is divided by metal strips into
squares that are a fairly intense "daylight" pink under
the directional fluorescent lights and a less intense
gray-blue under the cold-cathode lights that fill out
the rest of the ceiling. Even if this illumination were
the best possible one for viewing pictures, the pattern
of the ceiling, which suggests a vast, inane Mondrian
painting, seems too obtrusive in such low rooms. Before any more museums invest in similar lighting installations, the question of whether overhead light,
diffused and directed, is the answer to this problem
should be reviewed once more. Anyone who has
painted out-of-doors knows how the light from the
sky blots out the colors before him; one of the reasons colors are so brilliant in a camera obscura is that
the image is thrown on a table in a dark room, where
there is no overhead light to spoil the picture. Outdoors, to see true color values, one must either wear
an eyeshade or squint. Even when one is looking at
sculptures, which benefit by an overhead, natural light
(that is, the sun), it is better that the light fall at an
angle, so that the various planes of the modeling receive various intensities of illumination; an even diffused light obscures many details of the modeling.

In discussing this matter, I am happy to call to my
side a distinguished architect, Walter Gropius, and to
quote from his *Scope of Total Architecture*. He cites
a report by the Illuminating Engineering Society's
Committee on Art Gallery Lighting: "Today any interior (museum) gallery can be artificially lighted to
better effect than is possible by daylight, and, in addition, it can always reveal each item in its best aspect,

which is only a fleeting occurrence under natural lighting." Gropius comments, "A fleeting occurrence! Here, I believe, is the fallacy; for the best available artificial light, trying to bring out all the advantages of an exhibit is, nevertheless, static. It does not change. Natural light, as it changes continuously, is alive and dynamic. The 'fleeting occurrence' caused by the change of light is just what we need, for every object seen in the contrast of changing daylight gives a different impression each time." Those words, and the illustrations with which Gropius follows them, are so admirable and so right that I am tempted to add no comment except Amen.

There are, of course, buildings in which a uniform overhead light is aesthetically acceptable, as the new Manufacturers Trust Company building, on Fifth Avenue, has demonstrated, but even so, one may doubt whether uniformity of light does justice to the physiology of the eye and whether it does not, in fact, lead to loss of adaptability. A picture gallery is, at all events, not a bank. Apart from sterile geometric designs that might have been painted or might be best seen under artificial light, paintings, too, suffer under a uniform overhead light; even as sophisticated a spatter of paint as Jackson Pollock's "Number 27," in the Whitney Collection, loses some of its charm of texture and color under such illumination. Lighting engineers, who are proud of their new technical devices and who, coincidentally, have something to sell, may make it hard for an architect to resist their apparent authority as specialists. But lighting engineers have no standing as appreciators of art, and no mere photometer can ever record the only data of importance in a museum—the effect of the immediate environment and the image upon a responsive onlooker. So it seems to me that the extensive use of an artificial system of lighting in a building like the Whitney Museum, which could easily have had natural lighting on three sides (north, south, and east), registers a failure to make use of a fine opportunity. Only the galleries on the west side would have needed artificial lighting during the day, and they could have been reserved for the pictures that would lose least thereby.

Still, one must not be too severe on the designers of the Whitney for succumbing to the lure of technical overelaboration, since the fact that its galleries are air-conditioned offered a special reason for doing away with windows and the concomitant heat losses. Moreover, one must be endowed with truly heroic courage today to stand up against the fashionable trend toward supermechanization. The best corrective I know for the tendency to let mechanization take command of architectural design would be to heed what Tolstoy has to say toward the end of his *What Is Art?* Modern civilization, he points out, carefully seals up the windows and doors of a house and pumps the air out of it; then, observing with alarm that the people in the dwelling are being asphyxiated, it devises an elaborate and costly apparatus for pumping fresh air into it again, though all the poor inhabitants of it needed in the first place was a little natural air and light, supplied free by the simple device of opening up the windows. This applies to more functions than ventilation. In the next phase of modern architecture, perhaps, that little lesson will be taken to heart.

                                                                        1955

THE ROARING
TRAFFIC'S BOOM

TWENTY-THREE

Is New York Expendable?

The frantic effort to crowd the central district of
Manhattan with enough tall office buildings to make
traffic a permanent tangle is rapidly approaching complete success. Already, after ten in the morning, a reasonably healthy pedestrian can get across town faster
than the most skillful taxi-driver. All this may persuade someone in authority to suggest turning the
midtown district into a vast pedestrian mall, closed
to private vehicles during the day, as some of the
narrow streets in the financial district are now. Unfortunately, the load of pedestrians has likewise become so heavy, not merely at the lunch hour or during Christmas shopping but during most of the day,
that the walker is frequently slowed down to the ex-
hausting creep of the car or the bus. One would think
that this situation might cause some serious thought
among the bankers and investors and business enterprisers who have been fostering this congestion, admittedly with the sanction of the municipality's
zoning laws. Their lack of concern for the end product has been explained to me by one of the most
successful of our urban space men. " 'Money,' " he
said, giving the word the sort of halo a Roman might
attach to his tutelary deity, "is not interested in looking further ahead than the next five years." If this
truly represents the prevailing mood, the people who
are so ebulliently strangling the economic life of New

York and canceling out, one by one, every sound reason for living here must consider that New York is expendable. That is perhaps the attitude back of the recent proposals to build new super-skyscrapers on the sites occupied by Grand Central and Pennsylvania Station.

Before the plans for these buildings are rushed through, it might be profitable to consider just what is involved in these two projects. If one waits till they are built, it will be as useless to criticize them as it would be to reassess the wisdom of relying upon nuclear weapons after the cities of the world had been laid in ruins. Another point to consider is that the Municipal Art Society has designated both these stations buildings of architectural merit, and therefore worthy of preservation. True, the Society has put them in Category 3, which embraces "structures of importance . . . designated for protection," and not in Category 1, which embraces structures of national importance that "should be preserved at all costs." But if, in the Society's estimation, such buildings as the Morgan Library and other purely antiquarian older monuments belong in Category 1, both railway stations deserve such national recognition even more. The more venerable they become, the less significant their aesthetic mistakes seem. The Pennsylvania Station was conceived with a reverential view to reproducing the imposing masonry columns and vaults of Roman baths, and the façade of Grand Central is a decorative pastiche of Renaissance forms, but that does not nullify their real architectural virtues. And though the spacing of the columns at the Seventh Avenue entrance to Pennsylvania Station presents a formidable obstacle to automobiles going down the ramps from the street level to the main level, and though the multiple exits from the trains are a vexation to people trying to meet an incoming passenger, this station, like Grand Central, has positive qualities that make up for these functional errors. The major quality of each station, one that too few buildings in this city today possess, is space—space generously, even nobly handled. These two large gifts of space, though now broken by booths and sullied by huge advertisements, still give the tired traveler a lift when he leaves the comparatively cramped quarters of a train after a long trip. The combination of mass and volume is one of the special blessings of monumental architecture, and people journey thousands of miles to behold it in the remains of the Baths of Caracalla and the Colosseum. For this blessing, the hardheaded businessmen who fathered these stations were willing, in their own time, to pay—though it is likely that such men as Alexander Johnston Cassatt, then head of the Pennsylvania Railroad, knew full well the value of a penny.

Now, spatial magnificence cannot be justified on purely commercial or utilitarian grounds; it is one of those luxuries that pay off only through the centuries, by giving many generations the solace and delight of great art. The lofty ticket hall and concourse of Grand Central could be replaced by one a third as high, and a modern engineer could pump in enough air to fill the lungs of the crowds circulating through it, and the space saved by that operation could no doubt be profitably rented. But the aesthetic impression every user receives from a building modeled to delight the eye and not merely to accommodate battered humans would disappear. Anyone can verify this for himself by comparing his sensations at even a crowded hour in Grand Central or Pennsylvania Station with those stimulated by the Port Authority's low-ceiling bus terminal or with the even more oppressive subcellar effect of the Pennsylvania's Broad Street Suburban Station in Philadelphia, which is buried beneath a skyscraper.

In addition, Grand Central has the special merit of being so logically organized, so well handled in its details, that despite the mistake, back in the 'twenties, of allowing the Graybar Building, which flanks it on Lexington Avenue, to pre-empt space that could have been used for additions to the facilities of the terminal, it remains, forty years later, a masterpiece of both the architect's and the planner's art as practiced in the first decade of the century. From the beginning, these stations might have been as cluttered and debauched as they have since become (and, by the same token, ready candidates for replacement by purely commercial structures), for once steel-frame construction was invented, the railroads could easily have erected low, utilitarian station entrances and concourses and surmounted them with towering office buildings or hotels. After all, station hotels were already common in Europe, and since the trains were being put underground and propelled electrically, the elimination of smoke and noise bolstered the argument for the maximum use of the sites of Grand Central and Pennsylvania Station. But no such high-rent structures were included in the plans for them, perhaps because railroad management of that day had a more than vestigial sense of the responsibilities of great wealth and power—the sort of conduct that governed him whom Aristotle called the "magnificent man." "Money" was even then mightily preoccupied with making more money, but it was occasionally willing to look more than five years ahead. So much for the aesthetic and historic reasons for keeping these splendid stations intact as mementos of an era when our country was not too grubby-minded to believe that it could afford the luxuries of space and dignity and order.

By now, both stations need a thorough housecleaning, and Grand Central, with its tremendous litter of booths and advertisements, needs a visit from the junkman, too. But they still retain some of the essentials of great architecture, and many, many people besides the two hundred and twenty architects who recently made a public plea for preserving these terminals recognize the fact. In response to this plea, a scheme has been devised by the architectural firm of Emery Roth & Sons to keep the concourse of the terminal and replace the rest of it with a sixty-story building. Those who finally decide the fate of the terminal may accept some such compromise. But though this softens the charge of aesthetic vandalism, it does not lessen the proposed overcrowding. Because of the heavy press of human beings that crowds the midtown area, which is itself the product of planned congestion, the Grand Central concourse has become, in addition to its proper function, a secondary traffic artery. Many pedestrians walk east or west through the station because it is a short cut uncluttered by vehicles, and the morning and evening commuter rushes already choke the concourse. To pile twenty or thirty thousand more people in offices on top of the present facilities, to say nothing of the thousands of visitors to those offices, would make the station's daily pedestrian load what it is now only on certain frazzling national holidays. In addition, the secondary traffic needed to service all these extra people—deliveries to restaurants and the movement of other supplies and services—would so choke the streets around Grand Central that travelers reaching it by car or taxi would undergo more anxiety and miss far more trains than they do today. The only

way to make such a project "work" would be to stop
using it as a railway station. The latest scheme—that
of turning all the Grand Central properties over to
the real-estate firm of Webb & Knapp for development (read "greater building densities")—only adds
to the weight of this criticism.

It is incredible that the management of the railroads concerned should think of adding to their considerable problems by intensifying the congestion
around their most important stations; instead, they
should be exerting pressure upon the municipality
to halt the competition among builders to make the
whole midtown area untenantable. Constructing facilities that can only partly relieve the congestion
already created will cost the city many times the increase of tax revenue from the new skyscrapers that
will augment that congestion. Not only that, but the
delay and lost motion in street transportation is cost-
ing the community hundreds of millions of dollars a
year and steadily driving business to the suburbs. (As
far back as 1931 the Russell Sage Regional Plan cal-
culated that this loss came to $500,000 a day in New
York alone and to $1,000,000 a day in the metro-
politan area. These figures would now have to be
multiplied again and again.) If the city's railroads
intend to continue functioning as railroads, they
should look further than five years ahead and realize that unless traffic can circulate freely in the midtown area they will be among the heaviest losers.
What they are putting forth as an improvement is in
fact a final debasement, and what they propose as
profitable enterprise to make use of what they consider wasted air space will hasten the general metropolitan debacle whose symptoms are already obvious.
This is all the more true because in the act of creating more office buildings, the Grand Central "redevelopment" scheme may destroy essential hotel

accommodations and accelerate the pushing of hotels
for transients from the area in which they can serve
their patrons best with the least demoralization of
traffic.

Even if midtown business building were to halt for
a dozen years (it did halt almost completely from
1933 to 1947), that would hardly give time enough
to relieve the congestion that already exists. As for
most of the plans for improving the situation, they
seem to be the product of sleepwalkers who have
never observed the city by day. It would take a great
mind indeed to decide which set of planners is more
irrational—the people who are piling up high structures in the overcrowded business districts of our
cities, or the people who are creating cross-country
expressways that dump more traffic into them. The
Port of New York Authority and the Triborough
Bridge and Tunnel Authority have just come up with
a series of proposals for spanning the Narrows (thus
making Brooklyn and Long Island continuous with
New Jersey), for adding a six-lane deck to the
George Washington Bridge, and for expanding the
approaches to the Jersey side of this bridge—one of
those vast spaghetti messes of roads and clover cross-
ings and viaducts that provide excellent material for
aerial photography but obliterate the towns they pass
through as mercilessly as a new Catskill reservoir. The
assertion by the *Times* that these proposals "will
contribute only modestly to the relief of New York
City's acute traffic congestion" is a staggering under-
statement. It is also proposed to connect the George
Washington Bridge, via the Cross Bronx Expressway,
to the projected Throgs Neck Bridge. These schemes
give not the faintest thought to the immediate results
(piling more vehicles into an already swamped city)
or the end product (a massive influx of population
and industries into an area where the pressure of

both should be lowered, not raised). This will not
benefit even those who look only five years ahead.

Unfortunately, the promotion of traffic is itself a
big industry that gives play to all sorts of technical
and administrative skill. Then, too, the new bridge
facilities to be provided by the Triborough and Port
Authorities, which are part of this industry, can easily
pay for themselves through toll charges; and expressways and highroads enjoy the aid of state and federal
subsidies. Thus these ill-considered ventures, estimated to cost six hundred million dollars, present no
real problem of financing to delay their execution.
Meanwhile, New York's traffic commissioner, Mr.
T. T. Wiley, points out that neither of the Authorities is accepting the responsibility for disposing of
the new traffic that these bridges and highways can
pour into Manhattan. Leaving a baby on the doorstep is a mild analogy for this process; the traffic authorities are dumping a whole orphanage on an overcrowded and bankrupt home. But then, ever since
the nineteen-twenties the municipal and state authorities have been plunging blindly from one grandiose
traffic scheme to another, without showing any striking understanding of the problems they were trying
to solve. Traffic specialists take it for granted that
the aim of good traffic planning is to give the maximum accessibility and the maximum facilities for
movement by wheeled vehicles. But the aim of sound
city planning is to achieve a healthy balance between
the myriad activities of a community; it does not as-
sume that private transportation has a prior claim on
every resource of the community or that it is quite
all right to make the city less and less habitable as
long as enough roads are built to permit people to
escape by car once a week—only to crawl back, worn
and defeated, on Sunday evening. Transportation—I

blush to utter a truism now so frequently ignored—is a means and not an end.

To cure our creeping paralysis, our community has committed its destiny to empiricists (alias specialists and experts) who have little insight into the needs of the modern city. Situations that were foreseen and correctly diagnosed a generation ago by a few perspicacious planners like Clarence Stein and the late Henry Wright have become too acute for any simple set of remedies and too complicated to be treated without first framing a sound common objective that will guide each operation from one logical step to the next and link them all together in a new urban pattern. Sooner or later we must face New York's situation as a whole and make a fresh start.

1955

TWENTY-FOUR

The Two-Way Flood

What happens in New York to the art of building is bound up with what happens to the city as a place to work and live in. If it ceases to be a milieu in which people can exist in reasonable contentment instead of as prisoners perpetually plotting to escape a concentration camp, it will be unprofitable to discuss its architectural achievements—buildings that occasionally cause people to hold their breath for a stabbing moment or that restore them to equilibrium by offering them a prospect of space and form joyfully mastered. For a whole generation, New York has become steadily more frustrating and tedious to move around in, more expensive to do business in, more unsatisfactory to raise children in, and more difficult to escape from for a holiday in the country. The subway rides grow longer and the commuting trains carry their passengers from more distant suburbs, until as much time is spent in transporting the human carcass as is gained by diminishing the work week. Because urban surface transportation often comes almost to a standstill, the cost of delivering anything to anyone is rising steeply and the futility of owning a car for any purpose but fleeing the city over the weekend is becoming clearer and clearer. Meanwhile, the distant dormitory areas of New York describe ever wider arcs. By 1975, the Regional Plan Association's experts calculate, more people will be living in the suburbs within fifty miles of

New York than will live in the city itself. When that happens, it will be impossible to build enough highways to accommodate the weekend exodus, just as it is already impossible to provide enough internal traffic arteries to handle Manhattan's present congestion. And obviously, even if people could escape, they would then have no place within easy distance to go, since there would be no choice for recreation but metropolitan jam or suburban jelly.

Fifty years ago, the upper-income groups here, as in most other big towns, began to move out of the city along the railroad lines, to provide their families with peace and quiet, open spaces and gardens, and tree-lined roads that brought them quickly into the country for a walk or a picnic. "Life with Father" took on a rural tinge, though Father rarely got home in time to do more than say good night to the children. Since then, the desire to escape the city has filtered down into every other economic group, and as a result of the suburb's popularity in satisfying this desire, that haven of refuge is itself filling up. Despite village zoning laws, skyscraper apartments overtop the trees in regions that were rural only yesterday, and the load of metropolitan traffic on the parks and highways around New York, abetted by the subdivider busily turning farms into building lots, has enormously cut down the open spaces that gave the suburb, despite its inconveniences, an edge over the city.

Were the eruption of vehicles and buildings in and around New York a natural phenomenon, like Vesuvius, there would be little use discussing it; lava inexorably carves its own channels through the landscape. But the things that spoil life in New York and its environs were all made by men, and can be changed by men as soon as they are willing to change their minds. Most of our contributions to planned chaos are caused by private greed and public miscalculation rather than irrational willfulness. During the nineteenth century, when more cities were built than ever before, the business of assembling them was entirely in the hands of those who were thinking only of their immediate needs or their immediate profit. "Officers and all do seek their own gain, but for the wealth of the Commons not one taketh pain," a late-medieval poet commented at the very beginning of this urban breakdown. By now, hundreds of millions of dollars are poured every year into clearly obsolete and ineffectual efforts to overcome the ensuing congestion—street widenings, double-deck bridges, cloverleaf intersections, subways, garages—and the sole result of these improvements is to accelerate the disorder they are supposed to alleviate. Manhattan will soon be in the same predicament as imperial Rome; it will have to banish private wheeled traffic from the midtown area in daytime, as Julius Caesar did in Rome, to permit a modicum of public transportation and pedestrian movement. This will mean, as in Rome, the delivery of goods by night. That may temporarily relieve the congestion, but it will permanently increase insomnia, as Juvenal sardonically noted after Caesar issued his traffic ordinance.

I have put the whole urban picture within this ample frame to counteract the current habit of looking at one small corner of the problem—congestion at the Jersey end of the George Washington Bridge at 8:45 A.M. or at the intersection of Seventh Avenue and Thirty-fifth Street at 3 P.M.—and attempting to solve that. Perhaps the wisest words on the complexity of the traffic problem were uttered long ago by Benton Mac-Kaye, who fathered the Appalachian Trail. To relieve the congestion of traffic in Times Square, he remarked, it might be necessary to reroute the flow of wheat through the Atlantic ports. But our one-eyed specialists continue to concoct grandiose plans for highway development, as if motor transportation existed in a social vacuum, and as if New York were a mere passageway or terminal for vehicles, with no good reasons of its own for existence. To these experts, a successful solution of the traffic problem consists of building more roads, bridges, and tunnels so that more motorcars may travel more quickly to more remote destinations in more chaotic communities, from which more roads will be built so that more motorists may escape from these newly soiled and clotted environments. If these planners realized that it is as much the concern of good planning to prevent traffic from going into areas that should remain secluded and stable as it is to bring new traffic into areas that should be developed, they would never have offered their recent proposal for undermining what is left of rural Long Island. About that particular outrage, I shall have more to say in a later article.

The fact is that motor transportation is the sacred cow of the American religion of technology, and in the service of this curious religion no sacrifice in daily living, no extravagance of public expenditure, appears too great. Motor transportation is not merely an object of public worship; it has succeeded the railroad as the most powerful tool for either distributing or congesting the population—and it currently does both. Like any other tool, it must be used for some human purpose beyond the employment of the tool itself, and that further purpose represents the difference between carving and mere whittling. Our transportation experts are only expert whittlers, and the proof of it is that their end product is not a new urban form but a scattered mass of human shavings. Instead of curing congestion, they widen chaos.

The best way to understand what has to be done to make both the city itself and the surrounding region livable and workable is to begin our exploration of the problem at the center of the town and work outward to the country. The same principles of modern planning apply to both areas, and are just as difficult to put into action. The problem affects the working of every organ of the city, and neither gentle poultices nor brutal surgery (like the latest roadway encroachments on Central Park) will restore the city to health. The gridiron plan of New York, with its standard block, two hundred feet wide by six hundred feet long, and its numerous intersections, is a product of the age of the public stagecoach, the private carriage, and the common cart. The massive network of streets and avenues that, back in 1811, the City Planning Commissioners projected up as far as 155th Street was more than adequate for its original job. In fact, it was actually wasteful of land in residential areas, for it gave these quarters the same vehicular space it gave the busiest commercial districts, and thus sacrificed land that should have been dedicated to squares and parks and schools. But as long as only horses and carriages used the streets and only a fraction of the population could afford a private turnout, the streets of New York —at least above Washington Square—met the demands of transportation. This is not to say that there were not occasional traffic snarls and bottlenecks. For the safety of the pedestrian, the municipality in 1867 built a four-pronged footbridge over Broadway at Fulton Street, and if laziness had not prevailed over prudence, it might have built a series of footways over every busy intersection, like those that now span the East River Drive.

What caused traffic to become a serious problem was first the development of the multi-story building after the eighteen-eighties and then the rapid spread of the motorcar after 1915. There were forty persons per motor vehicle in the United States in 1915, a little over ten in 1920, and about five in 1925, and though at first New York didn't go in for automobile owner- ship as wholeheartedly as other parts of the country, its density of population produced the same problems here just as quickly. During the Second World War, owners were permitted to use the public streets gratis all week long for parking, and since then the number of cars quartered along our curbs has grown year by year. Our streets and avenues were designed to serve a density of population no greater than could be ac- commodated in buildings four stories high. But in a large part of Manhattan we have overlaid the land with so many high buildings that we have in effect piled from three to ten early Manhattans on top of each other. If the average height of these buildings was only twelve stories, the roadway and sidewalks flanking them should, according to the original ratio, be two hundred feet wide, which is the entire width of the standard New York block. In other words, to alleviate the pressure of traffic, we should be tearing down all the existing buildings in certain areas instead of put- ting up still taller ones.

As the city grew, the elevateds and subways took some of the load of traffic off the streets and tempo- rarily stalled off strangulation of Manhattan's north and south traffic, but these facilities effected practically no improvement in its crosstown traffic. Because our sub- way lines have opened up new sections of the Bronx, Brooklyn, and Queens and are pouring the rising pop- ulation of these districts into Manhattan, these lines now ease the difficulties of uptown-and-downtown cir- culation here only during the non-rush hours. The present area of our streets and avenues cannot be in- creased without adding to our already monstrous de- ficiencies in public parks and playgrounds in residen- tial areas where the population densities have mounted to between two hundred and four hundred and fifty people an acre. After all, the most feasible means of expansion, the river and belt-line drives, have already been resorted to. The only way, therefore, of providing space for all the present-day transportation, short of resorting to even more elevated highways, and thus creating a hell of noise and shadowed buildings far worse than the one our "L"s produced, is to build multiple-level tunnels under every congested street and avenue—which, as Euclid used to say, is absurd, since even a fraction of this construction would land the municipality in irremediable bankruptcy. We could, however, make better use of the land in Manhattan by replanning our residential neighborhoods into great superblocks, with fewer streets and fewer intersections, and all but purely local traffic confined to wider arteries that run past but not through these neighborhoods. We have the beginnings of this kind of development in Stuyvesant Town and many of our public housing developments, but nowhere has it been done systematically enough to provide wider arteries for even cross-town traffic. No piecemeal improvements, however valuable in themselves, can take the place of a scheme that will consider the city and its problems as a whole, not as something to be patched up here and there while the rest of it goes to ruin.

Instead of the city's preventing the internal traffic flood at the place where it originates, traffic engineers wait till traffic has reached the flood crest and then build ditches and canals to carry it away. These ditches merely add to the congestion, since traffic, unlike rivers, flows in two directions, and the wider the new exit route, the more traffic flows into it. All the current plans for dealing with congestion are based on the assumption that it is a matter of highway engineering, not of comprehensive city and regional planning, and that the private motorcar has priority over every other means of transportation, no matter how expensive it is in comparison with public transportation, or how devastating its by-products. In most cities, current plans for "traffic relief" include adding central parking areas, often tunneling under public squares and parks, as in San Francisco's Union Square Park, or of building handsome garages, like the one Philadelphia's special commission has lately opened on Rittenhouse Square, but all these devices merely invite more traffic. Our Park Commissioner, Mr. Robert Moses, has happily resisted this kind of encroachment on our own parks, but he is convinced that more garages should be built in the core of congested areas or just outside them. And only the other day Mr. T. T. Wiley, our Traffic Commissioner, put forth a proposal to provide room for forty thousand cars in public parking lots and garages, mostly at the edge of the city—as if this invitation to fill the highways coming into New York would magically take a load off the streets within the city. Proposals for off-street parking space in new office buildings continue to pop up, too, and the practice of providing at least enough parking space for the higher executives is fast becoming standard practice. One of the latest suggestions, by a firm of local architects, offers a slight variant on Mr. Moses's scheme—a series of huge garages, up to six hundred feet long and seven floors tall, astride the city's river drives. This scheme, too, would halt many incoming cars at the edges of Manhattan, but it suggests no means for enabling the dismounted motorist to reach his urban destination by swift public transport. Even if it did, it would still have the common failing of traffic schemes—it would encourage more vehicles on overburdened arteries. People, it seems, find it hard to believe that the cure for congestion is not more facilities for congestion.

While only a quack would pretend to have a pat solution for this complicated problem, there is no reason that we should not explore alternatives to the course we have been so blindly following. As Mr. Robert Mitchell, one of our few intelligent traffic experts, lately remarked, when a municipal counselor in Philadelphia expressed shock at the modesty of the budget he proposed for further research, "What we need is not more traffic counts but more thought—and thinking is cheap." A little thought may disclose that with traffic, as with many other matters, there is no swift and simple answer—"the longest way round is the shortest way home." Before we can promise to restore the normal facilities of transportation to our blocked avenues and our almost paralyzed metropolis we may have to take even more drastic measures than rerouting the continental flow of wheat. Most of these measures, happily, will increase the habitability of the city and relieve the almost neurotic compulsion to get out of it. But one cannot promise that this public gain will produce a private profit.

1955

TWENTY-FIVE

Restored Circulation, Renewed Life

Most of the fancy cures that the experts have offered for New York's congestion are based on the innocent notion that the problem can be solved by increasing the capacity of the existing traffic routes, multiplying the number of ways of getting in and out of town, or providing more parking space for cars that should not have been lured into the city in the first place. Like the tailor's remedy for obesity—letting out the seams of the trousers and loosening the belt—this does nothing to curb the greedy appetites that have caused the fat to accumulate. The best recent book on the subject, *Urban Traffic,* by Robert B. Mitchell and Chester Rapkin, takes quite another view—that traffic is but one "function of land use," which is to say that streets and highways should not be treated as if they existed in a desert inhabited only by motorcars. How different that attitude is from the prevalent conception, as succinctly summarized by a one-time city-planning commissioner: "The main purpose of traffic (surely) is to enable a maximum number of citizens to derive all possible benefits from the use of automobiles as a means of transportation, for business, convenience, and pleasure." It is because this second conception of traffic is dominant that our cities have become a shambles.

Before we cut any more chunks out of our parks to make room for more automobiles or let another highway clover leaf unfold, we should look at the transformation that has taken place during the last thirty years in Manhattan—a city that is steadily growing higher, denser, more complex, more clotted, more confused, its chaos solidifying into an insane mess of high buildings placed within a rigid urban framework that is hopelessly out of date. Our mild legal limits on the height of midtown buildings merely encourage tall structures in the very areas where traffic congestion is already close to paralysis, and we demolish crowded slums only to replace them with public-housing developments whose population densities, as high as four hundred and fifty people an acre, are twice the average residential density of the city. We have consistently acted as if there were no relation between the number of people we dump on the land and the amount of congestion on the streets and arterial traffic routes.

Instead of maximizing facilities for motorcars, we should maximize the advantages of urban life. Parks, playgrounds, and schools, theaters, universities, and concert halls, to say nothing of a quiet night's sleep and a sunny outlook when one wakes up, are more important than any benefits to be derived from the constant use of the automobile. To accomplish this improvement, we must devise a fundamental change in the city's whole pattern. The plain fact is that the high-density city is obsolete. If the city is to become livable again, and if its traffic is to be reduced to demensions that can be handled, the city will have to bring all its powers to bear upon the problem of creating a new metropolitan pattern, not just unintegrated segments of such a pattern, like the dubious public-housing projects of the lower East Side.

A large part of the present difficulty (as visible, by the way, in London, a city of low buildings, as it is in New York) is caused by the over-employment of one method of transportation, the private motorcar—a method that happens to be, on the basis of the number of people it transports, by far the most wasteful of urban space. Because we have apparently decided that the private motorcar has a sacred right to go anywhere, halt anywhere, and remain anywhere as long as its owner chooses, we have neglected other means of transportation, and have even permitted some public mass-transportation facilities to lapse while our municipalities and states spend public moneys in astronomical amounts to provide additional facilities for private transportation. The major corrective for this crippling overspecialization is to redevelop now despised modes of circulation—public vehicles and private feet, both of which are essential for mass movement. An effective modern city plan would use each kind in its proper place and to its proper extent—the walker, the vertical elevator, the private car, public surface and subway transportation, and (for longer distances) the railroad, to mention them in the order of increasing speed and capacity. Only when all five are made use of and planned in relation to one another can an efficient circulation of traffic be maintained. When, for example, the vertical elevator is used to excess, it produces a mass of buildings so high that no feat of horizontal mass transportation can handle the resulting human traffic without insufferable rush-hour jams. On the other hand, if the jumble of enterprises that now clutter the great midtown and downtown areas of the city were more generally dispersed—in the way that Macy's has established a branch in Flatbush, Bloomingdale's has set up one in Fresh Meadows, and so on —this would take some of the burden off the congested central district, and many people who are now long-distance shoppers by car might become pedestrians again.

I have suggested earlier the possibility that private vehicles may eventually be excluded from whole urban areas. Do not fancy that this is a mere whimsey. It has already happened in the Wall Street district, and through traffic has for a long time been barred from Manhattan's "play streets." Philadelphia's director of city planning, Mr. Edmond Bacon, has suggested that its busiest shopping thoroughfare, Chestnut Street, might be turned, with profit to all concerned, into a pedestrian mall, forbidden to private motors or taxis. Fifth Avenue, during the hours when its traffic is stalled, is for all practical purposes now a pedestrian mall between Thirty-fourth and Fifty-ninth Streets. Removing the cars from it might make it as pleasant a place to shop in as Amsterdam's Kalverstraat or Buenos Aires' Calle Florida, both of which are sacred to pedestrians. The replanning of New York so that the pedestrian may again have a real place in the urban economy would have seemed fantastic only a generation ago. But if the pedestrian is to come back, it is necessary, for both his safety and his health, to insulate his promenades from the traffic thoroughfare, just as it is necessary to keep motorcars from entering areas where they do not belong and to provide for their swift movement through areas where they do belong. Concern for urban health should be no small incentive in this planning, for the increasing quantity of lethal carbon monoxide poured into the air by the internal-combustion motor has become a serious occupational hazard for traffic policemen, and doubtless in some degree lowers the vitality of everyone who uses the streets and breathes the foul exhausts.

The principle of separating walkers from drivers, which involves planning whole neighborhoods at a time, is known to planners as the Radburn idea, after the planned town of Radburn, New Jersey, laid out by Clarence Stein and Henry Wright, but as a matter of fact it was first embodied in the plan of medieval Venice, whose canals carried the swift-moving traffic of another age. Until Radburn was designed, in 1928, no professional planner seemed able to understand that the extraordinary charm of Venice, which persists despite its overcrowding and decay, is due partly to the fact that each neighborhood was planned as a unit, for the benefit of the foot walker, and is not menaced by the rumble and roar of wheeled traffic, and that to go from one part of the city to a distant part one uses an entirely different transportation system, which never suffers any interruption by the pedestrian and does not interrupt his progress, either. Leonardo da Vinci proposed to overcome the congestion of Milan by a similar separation of wheeled traffic from pedestrian walks. The first modern planners to effect such separation were Olmsted and Vaux, in their brilliant plan for Central Park. Their scheme provided separate ways for the pedestrian, the horseback rider, and the carriage driver—to say nothing of confining commercial traffic to the "expressway" transverses—and it minimized the number of intersections by using overpasses and underpasses. If you examine the original plan of Central Park, you are examining a modern city plan, and if you walk through the Mall, noting how the traffic circulates around it, you have only to imagine buildings spaced at intervals along it, in related groups, to understand the principle of the superblock, which should be the minimum unit of land subdivision for the ideal big city, as against the standard New York block, whose inadequate size is one of the chief handicaps to a sensible redevelopment of our metropolis. Harvard Yard, in Cambridge, is a superblock; indeed, Cambridge is full of mid-nineteenth-century superblocks, with economical cul-de-sacs (rather than the conventional space-wasting gridiron of streets) and spacious gardens that have proved a happy barrier to overcrowding. And Rockefeller Center, too, is a quasi-superblock, though by no means a perfect example, since the unifying pedestrian feature is underground. While it maintains the gridiron street pattern, and has, in fact, even added a north-and-south street in its middle, it has at least demonstrated that a related group of office buildings, with plenty of room for pedestrian traffic, has vast advantages over the average helter-skelter city block.

The superblock—a unified campus, or precinct, as the British call it—is now the fundamental unit of modern urban planning. Instead of fronting buildings on streets and stringing stores and offices along avenues, modern planning insulates wheeled traffic and groups related buildings into campuses and unified working quarters, scaled to the pedestrian, with every necessary utility or facility concentrated close at hand.

The superblock was not an invention of Olmsted, but he was the first to use it, back in the early 'nineties, for his communities of Riverside, Illinois, and Roland Park, Maryland. It permits the grouping of housing units around a decent minimum of open space, or even park, without the intrusion of through vehicular traffic, an intrusion that the standard city block, and our gridiron layout of streets, actually invites. Instead of sacrificing good living quarters to traffic, you improve both at the same time by insulating them from each other. Radburn was the first community to embody all these ideas for the modern city. This philanthropic venture of the farsighted realtor Alexander M. Bing and his City Housing Corporation was an early victim of the depression, and thus never became the complete town it was designed to be. But two whole neighborhood units, united by a system of pedestrian ways entirely separated from the motorways, were constructed. There are several methods of building such a town. Stein and Wright created a continuous core of park, around which the neighborhoods, separated from each other by through traffic arteries, were arranged. The planners of the English garden cities, on the other hand, used the park as a green belt to both surround the outer edges of the neighborhoods and set the neighborhoods apart. But wherever a new city is being built today, whether it is Kitimat, the new aluminum-making town in the wilds of British Columbia, or Chandigarh, the new capital of East Punjab, in India, the principles that Wright and Stein incorporated in Radburn, which was deliberately conceived as a "town for the motor age," are being employed. Even Le Corbusier, after suggesting many extravagant and often silly ideas for the city of the future, has belatedly adopted, without much modification, the Radburn plan.

Paradoxically, a "town for the motor age" does not sacrifice either beauty or habitability to the overriding needs of the motorcar. The system of planning cities as a group of neighborhoods and precincts, with internal traffic minimized and outside traffic excluded, can be applied to old cities, too. Indeed, if the centers of our big metropolises are not to be blighted by the competing attractions of suburban shopping centers and suburban business plazas, with their comparatively ample facilities for transportation and parking (the result of more open building and therefore lower densities of population), they can do it only by bringing about a change in their own plans. It will take time and money, of course, but in the end the wrong sort of planning takes much more time and money.

This new kind of city design actually meets the needs of both business and living much better than any plan that makes the transportation system the dominant element. There is a natural tendency in all cities for related trades and occupations to gather together in districts of their own, regardless of the traffic problems such concentrations of industry and people produce. The financial interests—banking, brokerage, and their necessary attendants, the lawyers—have long clustered along Broad and Wall Streets, and it was only lately that they set up a subcenter in and around Rockefeller Center. The garment industry clung, during the sweat-shop days, to the congested lower East Side, where the work was farmed out to the people who lived in the crowded tenements, and it now forms an even more congested city in itself in the West Thirties—one so crazily conceived that it takes longer to send a rush order to Fifth Avenue and Fifty-seventh Street than it should to send it to Philadelphia.

Obviously, no private agent, not even a wealthy life-insurance corporation, can undertake the task of regrouping the business, the industrial, and the residential areas of New York so as to facilitate their activities and reduce the amount of long-distance traffic in the city, for such a project requires not only immense sums of capital but the full use of public powers. At present, the municipality has no legal authorization to acquire by condemnation land for such large-scale redevelopment. The only agency that has public powers on a regional scale sufficient to deal with even the traffic aspect of the problem is the Port Authority. Unfortunately, this body has taken its privileges as a profit-making corporation more seriously than its public obligations as a planner of interstate industrial and civic enterprises, so instead of helping to solve our traffic problems, it is now one of the major interests fostering congestion, and it levies tolls on that congestion in every new tunnel or bridge it builds. To replace our present bungling palliatives for congestion with a plan for building a permanently attractive and livable city, the municipality needs legal powers to acquire and redevelop land for a variety of purposes. It should also be legally enabled to plan residential neighborhoods for more than merely the lower-income groups, and to plan business quarters on a far greater scale than Rockefeller Center.

In a city already as deeply committed to congestion as New York, with its property values, its system of taxation, its budgetary needs all geared to the policy of furthering congestion, genuine planning can make headway only gradually, and after much public education. In redesigning New York, we would, by the way, do well to heed the precedent, ignored by our planners, that was set long ago by Superintendent John Tildsley, who sited the public high schools erected in the city during his tenure so that the students would mainly travel against the prevalent streams of traffic morning and afternoon. Shopping centers and business quarters should be planned in the same fashion, to disperse some of the institutions of the city not to distant suburbs but to its own peripheral sections, instead of so concentrating them that it is necessary to pump almost four million people daily into the area south of Fifty-ninth Street.

If Manhattan were replanned in this new fashion, what would it look like? For another thirty years, it would hardly be possible to note the difference merely by gazing at the silhouette of the city; from a distance it would still be the same hilarious upheaval of steel and stone, romantic beyond words when viewed against the sunset from the approach to the Triborough Bridge or from Brooklyn Bridge. But in time the great volcanic palisade of buildings in downtown and midtown Manhattan would give way to whole quarters in which the tall structures—none of them, even the office buildings, over fifteen stories high—would be widely spaced and placed near the mass transportation routes and stops, while the trees and the grass-lined walks within these quarters would have the charm of a Paris boulevard, without the stench and noise that now make the café terraces of that city a humbug and an ordeal, as far as aesthetic pleasure goes. Each of these new quarters would have its own form and character, with its own social core of shops, markets, restaurants, churches, and schools, no longer scattered at random on through streets and avenues. Perhaps as many people as now enter the city from a distance by car would be able to take a salubrious walk to their work, just as the residents of the more fortunate areas of the East Side now do in the midtown area, though the salubrity of *their* walk is sadly tinctured by carbon monoxide. The great tides of traffic, instead of sluggishly moving in congested streams, would flow rapidly along new traffic routes, including crosstown expressways, and only vehicles that had business there would filter through the smaller capillaries into the unified neighborhoods. The daytime population of Manhattan would decrease, but as the city became more livable and the decayed neighborhoods above Fifty-ninth Street and below Twenty-third Street were restored to life, the living-in population might go up a little. One might again think of raising children in such a city without worrying about dope peddlers, juvenile delinquents, psychotics, and the dangers the young now encounter even on an afternoon walk in the public parks.

I do not suggest that the municipality can effect such a change while the whole force of government highway aid and business enterprise is addressed to promoting the growing congestion in the area both in and outside the city of New York. No internal corrective for congestion will work unless the principle of decentralization is applied to a much wider area. The kind of planning that stops at the limits of the metropolitan zone is as useless as that which stops at the legal limits of the municipality. Such planning forgets that there is a new, modern scale of distances, and that problems that were once soluble within a city now involve public control and development of enormously larger areas. We must not merely think of planning satellite communities around New York; we must overcome the highway engineers' itch to congeal into a solid urban mass towns and rural areas that should retain their individuality and their comparative independence. If this kind of long-range thinking gives anyone a headache, he is at liberty to exchange it for our present headache—a New York inhabited by a shifting, overcrowded, demoralized population, and a superhighway system jammed with people fleeing not from disaster but from the very city that is supposed to offer all the benefits that make life desirable.

1955

TWENTY-SIX

From the Ground Up

Practically all the impressive architecture and good planning in this country today are in low-density areas outside the big cities. It is there that the efficient horizontal factories are being built; it is there that the capacious shopping centers are being erected; it is there that the motorcar at last can translate speed into space, by giving small towns within a radius of thirty miles or so the social and business facilities that were once available only in congested centers. While every big city is beginning to feel the competition from the new suburban communities, imperfect and incomplete though they are, the lesson of their existence has still not been learned by those who are planning and rebuilding New York. The important point about this spontaneous decentralization is that, instead of trying to hold it back, the big city must try to direct it, to the advantage of both itself and the smaller centers. Furthermore, instead of attempting to pit the attractions of metropolitan overcrowding against the lure of the suburb, we must think of rebuilding the interior of the city, with gardens and parks and open vistas, so that it, too, will be desirable and habitable.

Unfortunately, there is a painful reluctance either to face the dire predicament of New York, with its mounting financial deficiencies and social delinquencies, or to discuss any alternative method of growth. The recent announcement by the real-estate firm of Webb &

Knapp that it intends to go ahead with its plans for another enormous building in the already super-con-
gested West Thirties only highlights this general failure
to be concerned with New York's survival. Yet by
now it should be plain that the present pattern of New
York's congestion is not merely costly, to the point of
municipal bankruptcy, but obsolete; the city, in fact,
has been rapidly strangling itself by the sort of building it has permitted. Because the costs of the fashionable dodges for relieving congestion vie with the
costs of congestion itself, the municipality now lacks
funds to provide proper schools for tens of thousands
of its children or to provide adequate pay for its policemen and firemen, to say nothing of coping with
the growing deficit in urban recreational space as
population densities increase. A unified policy of orderly, large-scale decentralization and rebuilding—
both within the city and in areas that lie far beyond its
present suburban limits—should be the principal basis
for overcoming the present old-fashioned pattern of
growth.

The case for decentralization can be demonstrated
by the experience of one of the city's most admirably
organized institutions, the Public Library. If the Library operated only in its Central Building, at Forty-
second Street and Fifth Avenue, its borrowers would
jam the transit lines in the neighborhood and would
have to queue up like people lining up for the movies
on Saturday night in order to make use, under har-
assing conditions, of its cramped facilities. The cure for
the city's overcrowding is the one the library long ago
adopted—not a bigger building in midtown, but a
series of branch libraries elsewhere, and a system of
distribution that links together the resources of the
whole organization. Other institutions, notably department stores and banks, have for some time been
following suit. I remember, though, that when I pro-

posed this measure, in 1929, to Mr. Jesse Straus, then
president of Macy's, he dismissed the suggestion as the
dream of a young idealist. Had he lived into the
'forties, he might have regretted his failure to buy good
sites for branch stores in other boroughs when land
there was cheap.

Acting on this principle of decentralization, the
municipality has increased the capacity of its public
colleges by planting offshoots in Brooklyn, Queens, and
the Bronx, and if it is wise it will meet the further demands that will tax these already too huge institutions
by creating still more independent units, since every
institution, as Aristotle observed, has a limit of size,
like "plants, animals, implements; for none of these
retain their natural power when they are too large or
too small, but they lose wholly their nature or are
spoiled." At a certain stage, there is a choice between
swelling an institution to the bursting point or dividing
and reorganizing it into a series of related units or cells,
limited in size, partly independent and self-contained,
but greatly strengthened because of their ability to call
for aid and sustenance from the parent institution. New
York, as a municipality, has not made this choice; indeed, it is talking of rebuilding the Washington Street
Wholesale Market—whose costly inefficiency adds to
the price of perishable foods in the whole city and its
environs—in its present cramped location, amid congested traffic, instead of removing it to a more suitable
area, related to the wholesale markets in the other boroughs. Since our officials shrink from organic remedies,
we continue to produce dangerous clots of congestion
—arterial thrombosis, in fact—at the center and an
undirected spill of population into the suburban areas,
and this, as Mayor Wagner mournfully observed the
other day, takes away the income groups that pay the
highest taxes.

The failure to change the pattern of congestion

within the metropolis has been aggravated by the reckless subsidized building of superhighways outside it; this has pumped more cars into the city and populated nearby open spaces that should have been permanently preserved, in part, as rural areas. Highway building on the lines now fashionable has steadily added to the disorganization and frustration of urban life, for it increases random movement without creating a new pattern of urban settlement. Its chief result is to extend the congestion of midtown New York to outlying areas. One of the superstitions of traffic engineers and city planners is that the aim of a good highway system is to increase the scope, penetration, and use of the motorcar—to the exclusion of every other form of locomotion and the sacrifice of every other urban function. In their preference for this mode of transportation, they *seem* at present to have the backing of the majority of the American people, who remain strangely quiet and passive about the matters that should concern them most. But if the motorcar is most efficient in the open country, it is least so in a city, and though the prospect of speeding at fifty or sixty miles an hour through the open country is what lures people out of New York on a weekend, it is not at this rate that the first twenty miles of the journey are achieved. It is only by creating a widely diffused network of roads at the same time that we create an equally decentralized pattern of urban settlement in communities surrounded and protected by permanent belts of open country that the advantages of motor transportation can be maintained. For the New Yorker, therefore, the possibility of getting out of the city swiftly and pleasantly for a day's or a weekend's outing depends upon the city's being set in the midst of a region that is still mainly rural. And since more and more people, and their motorcars, are now piling up in the outlying areas as permanent residents, our weekend efforts at escape are becoming more and more difficult and depressing. Yet public funds for housing mortgages and highway building continue to encourage the real-estate speculator and the superhighway builder to destroy vast adjacent rural areas of land that should be preserved not only for the now dispossessed market gardener, farmer, and chicken raiser but even more for the benefit of New York's millions.

Highway engineers currently act on the principle of the hostess who, spying at opposite ends of a crowded drawing room two people who have not yet met, thinks only of how to bring them together, though in doing so she may jostle and squeeze against her other guests, interrupt conversations, knock the cocktail tray out of the butler's hands, and embarrass the two recipients of her intentions, who would have been far happier had they been left alone. Take the proposed new highways to connect Long Island at two points with the industrial hinterland of New Jersey. These two direct express routes will expose the western part of Long Island to a flow of motorists from northern New Jersey, taxing recreational resources that are now, on a summer's day, already overcrowded. Moreover, the impending highway bridge over the Narrows, to connect Staten Island and Brooklyn, will complete the destruction of Staten Island as a natural recreation area for the surrounding region—a destruction that was set in motion by the building of the Kill van Kull bridge, connecting Staten Island with New Jersey. Now the complete industrialization of Staten Island and a far more intensive development of the Bay Ridge area of Brooklyn, two of the finest living and recreation areas in the five boroughs, seem inevitable.

Today you might easily forget—so enormous is the damage done—that one of the major assets of New York as a city was its nearness to the equable climate and the other natural resources of Long Island. This "fish-shaped Paumanok," with its barrier beaches and its salt marshes and its dunes, was one of the last great refuges of wild life near New York. Its inlets and bays harbor as fine clams, oysters, and scallops as can be found on the Atlantic coast (though by now they have been overexploited); and its farms, which up to 1895 provided a large part of New York's milk, still come through with marvelous crops of white Peking ducks and equally white cauliflowers, to say nothing of fine early potatoes. Rural and coastal Long Island are as much needed as a breathing space for the folk on Manhattan as is Central Park. So long as Long Island was effectually insulated, with only one bridge to join it to the mainland, with no motorways and only the worst railroad in the world to knit together its isolated communities, it maintained a healthy rural economy, protected from the encroachments of the city by a wide green belt of big estates. Even before the income tax made life difficult for owners of these country seats Mr. Robert Moses's magnificent parkways ruthlessly broke into the private preserves and opened the way for a new kind of development. And yet these highways, along with such big public recreational areas as Jones Beach, might have done much good if the state had adopted a policy of keeping the rest of the island as a mainly rural and recreational refuge. If public bodies had acquired strategic buffers and strips of land along the shores and at intervals across the island, to take the place of the private estates, the creeping blight of realty speculation might have been controlled. It is not too late now, perhaps, to save the eastern half of the island. But instead of preserving the remnants of Long Island's isolation, the highway authorities, with no thought of either urban New York's or rural Long Island's needs, now plan to give direct access to this area from heavily populated industrial New Jersey. Thus we can look forward to a series of

Newarks, Levittowns, and Coney Islands clear out to
Montauk Point, and a generation from now, East
Hampton may be indistinguishable from Canarsie.
What will have been lost? Only New York's greatest
natural asset—an accessible rural resort and an abun-
dant larder of succulent food. What will have been
gained? Ask the highway engineers and the realty
operators.

Our city's very existence is dependent upon maintaining the areas around it in a state of healthy balance, so that its supply of water will be safeguarded
and those who need the contrast and relief of a day or
a weekend in the open country—a privilege theoreti-
cally available to everyone now by reason of the shorter
work-week—may enjoy these things without losing
half the pleasure through the effort needed to reach
their destination. During the past four years, the population of the suburbs and smaller towns around New
York has grown four times as fast as the population of
the city. The effect of this is to concentrate an ever
heavier load of weekday traffic upon the highways
pointed toward New York and to add an ever larger
part of the suburban population to the hordes escaping
from New York on weekends.

If the rape of Long Island does not concern you,
consider Westchester, the only nearby countryside that
vies with its sandy rival. The people who built suburban homes there fifty years ago moved to a county
still almost as rural as Dutchess County is today. Their
children were the beneficiaries of the wooded ridges
and the pleasant streams that ran near every suburb.
Moreover, back in 1915, the first of the great parkways, the Bronx River Parkway, was built *not* pri-
marily to facilitate traffic but to conserve the Bronx
River Valley, too often used as a convenient refuse
dump, in accordance with the strange fashion Ameri-
cans have of picking out the best pieces of scenery for

that purpose. This long strip of parkland was the first of a series of green belts and playgrounds designed by Westchester officials for the people of the surrounding communities. Some of this land was left wild—Saxon Woods Park, Blue Mountain, and Pound Ridge reservations—and some of it, like the original Bronx River Parkway and the Hutchinson River Parkway, was handsomely landscaped, and these lovely green areas were used by families for hikes and picnics. Naturally, this open space attracted immense numbers of people from New York. But as congestion grows in the city, it grows on the highroads; by now, the traffic is so heavy that the strips of parkway bordering these arteries are being carved up to make expressways that it is hoped will speed up the gluey crawl of the Sunday driver in his search for open country. So the suburbanite who twenty years ago had the country at his doorstep must also pile his family into the car and seek it farther away. Swath by swath, these restful green belts are turning into speedways droning with traffic, while local woodlands, golf courses, and hiking grounds, offering the pleasures that drew the suburbanite to the suburbs in the first place, are disappearing, too.

Having replanned this whole rural area to make it unsafe and unattractive for the pedestrian, the highway engineers are using the resulting "way of life," which depends upon the constant use of the motorcar, as justification for more depredations on the landscape, creating miles of desert in which only the concrete cloverleaf blooms. In the suburb as in the crowded city, land values have risen as living values have gone down. This is the last step in what might be called the cycle of environmental impoverishment—i.e., metropolitan congestion and physical frustration; suburban escape; population pressure; overcrowding; extravagant highway building to promote further channels of escape at greater distances from the once so admirable center;

finally, intensified congestion both in the original center and in the suburb, which wipes out the social assets of the city and the rural assets of the country.

Meanwhile, water famines threaten every metropolitan area, and big cities like Philadelphia and New York compete for the use of the same river systems. By no effort of its own can New York safeguard its water supply, its recreational area, its local food supply, its access to the wilderness. All these facilities lie outside its political domain. What the city needs is a public policy that has in view the development of a much greater region than the metropolitan area. New York City cannot by itself handle the problems its own expansion and congestion have produced. All the technical dodges for curing the evils of overgrowth have the perverse effect of aggravating the disease. The way out demands political intelligence operating at every level—industrial, financial, political, cultural—so that a new pattern of urban development will be brought into existence which will create a balance between city and country, between centralization and dispersal.

Thirty years ago, the situation was analyzed by a group of men—architects, city planners, economists, educators, and foresters—and their remarkably accurate prognosis was published in the *Survey Graphic* in 1925. Fortunately, two of the leaders of this group were officials of the New York State Housing and Regional Planning Commission—Clarence S. Stein, its chairman, and the late Henry Wright, its leading planner. In 1926, this commission's final report, the first report of its kind to be issued in this country, traced the growth of the Empire State and its cities through two main stages, and indicated that a third stage was developing. In the first epoch, one of highways for horse-and-wagon transport, water mills as the main source of power, and canals, the population of the state was distributed quite evenly, with only a slight thickening at the chief ports—New York, Albany, and Buffalo. Then, after 1880, there followed a period of concentration, brought about by the railroad. Food and power were now transported from distant farms and coal mines, and the big city, offering a surplus labor market of new immigrants and a multitude of new opportunities for gain, grew at the expense of the smaller town. As a result, population became swollen in New York and Buffalo and shrank in other areas, while urban growth in general followed the railroad lines.

By 1925, when this report was drafted, a third stage was already plainly discernible: because of the development of electric power and the motor road, the distribution of power and other resources was no longer tied to either the railroads or the great ports. Technically—but not, alas, financially—the pattern of congestion was becoming obsolete, and these new facilities offered the first effective way of relieving the apparently almost unavoidable congestion of New York. This course of development, Henry Wright pointed out, did not favor a return to the original scattered distribution of industry in villages and small towns; it called for building up a series of self-contained industrial communities, spaced at intervals over a far wider area than the largest metropolitan area but closely interrelated. As for the increased population, Wright remarked that it was now practical to take care of it, too, in balanced communities of moderate size, set in the midst of permanent countryside. The place for these communities and industries was not mainly around New York, where they are presently being established, but in a belt settlement, made feasible by the motor highway and the electric-power line —a belt five or six hundred miles long and fifty to a hundred miles wide. The survey indicated that the best solution was to use a fertile band, narrow along the Hudson, broad along the Mohawk Valley and around the western part of Lake Ontario, where the climate was more pleasant than in Buffalo. With such a diffusion of population, highways could be planned to lessen congestion within the New York area and promote swifter, safer movement outside it. By public reservations and the kind of land-use policy later established in England—which makes it impossible to intensify the use of land without justifying such use before the proper state authority—the whole area around New York City, instead of being eaten up progressively by sprawling suburbs, would be punctuated by permanent belts and wedges of open country.

Two parts of this general regional transformation suggested have already been carried through: the conversion of the marginal farmlands of the Adirondacks and the Catskills into state parks, and the project for the St. Lawrence River seaway. The present Thruway from Buffalo might be useful as the spine of a new system of decentralization if, instead of pumping traffic into the heart of New York, it sought to promote lateral urban growth on each side of its route, and stopped short of the metropolis, perhaps opposite Tarrytown. The need for establishing a regional pattern of urban growth, which should have been one of the main motives of the Thruway, has been belatedly recognized by a private corporation, the City Investing Company, which has put forward a plan for a new industrial community, called Sterling Forest, on a seventeen-thousand-acre tract near Tuxedo, New York, alongside the Thruway. What is important about this new project is that, though the site is only thirty-one miles from the George Washington Bridge, it is planned not as a suburb but as a balanced industrial community, exploiting its own resources—peat humus, clay, limestone, and iron—and attracting a variety of industries and research laboratories. This is an exemplary move, but unless such private enterprises are protected by a farsighted public planning authority that has jurisdiction over a much wider area, they are likely, if they do not fall by the wayside, to merge into the clotted urban mass that is spreading outward from the city.

But apart from such projects as Sterling Forest, still in the paper stage, the growth that stubbornly continues to take place moves mainly along obsolete lines. This growth ignores the advantages of a decentralized pattern and the terrible liabilities of the congestion that results from funneling every activity into a single center. Actually, the decentralizing forces and techniques have got even stronger during the last thirty years than Wright and Stein and their colleagues dared to predict. People in the smaller communities may see the same television shows, listen to the same radio stations, read the same newspaper on the same day, go to the same motion pictures as people in the big city, and because of the low density of population they may use the motorcar with a sense of freedom and pleasure no longer possible in the metropolitan district. But the decentralization has been halfhearted and random, securing only the minimum enjoyment of its potentialities. Plainly, the continued dumping of more and more people into suburban subdivisions, the overcrowding of the once isolated country towns, like White Plains and Tarrytown, and all the various other losses and depletions of the rural landscape I have just been describing do not produce a desirable habitat. This sort of automatic pseudo-decentralization only adds to the area of confusion and chaos.

The kind of thinking that went into the preparation of the 1926 report on a plan for the State of New York, and into Clarence Stein's recent book, "Toward New Towns for America," is the kind that sends Mr. Robert Moses and other supposedly hardheaded administrators into apoplexy. Mr. Moses uses the word "regional planning" as a swearword, to indicate his abiding hatred of such comprehensive and forward-looking policies, just as he invokes the term "long-haired planner" to designate anyone who turns up with a proposal that does not fit into his own set of assumptions, most of them by now manifestly inadequate and badly out of date. Just the other day, nevertheless, Mr. Moses admitted in the *Times* that "traffic control" is "another of our few failures." Did he have the grace to add *"Mea culpa!"* under his breath? Certainly it is not Messrs. Stein and Wright who look silly now that all the predictions they made about New York's traffic congestion and the fashionable remedies for it have been verified a dozen times over. The fact is that their plan for the State of New York, though put forth only in outline thirty years ago, is decades ahead of any of the past activities or present proposals for curing traffic congestion or overcoming the immense disorganization and miscarriage of life that is taking place. Meanwhile, the present processes and plans keep on working auto-matically, and we go on, like so many busy beavers in flood areas, building new suburbs that will be inun-dated, only to build the same kind of suburbs a few years later a little farther away. Meanwhile, likewise, more dismaying congestions heap up within the city, and the only prospect before us under the present system of planning and building is "more and more of worse and worse." Even those who think only of feathering their own nest are actually soiling it so badly that it will soon be uninhabitable. The hopeful alter-native is not an idle dream, but it will require an ability to face realities, a sense of public responsibility, and a boldness of imagination that have so far been absent among those who have exercised authority. The I.B.M. Corporation, one of our leading practitioners of de-centralization, should distribute its favorite motto to our planning authorities, our highway engineers, our banks and insurance companies, our real-estate developers and, indeed, to our whole body of citizens and voters. *Think!*

1955